HOW DID THEY DO THAT?

D1018783

HOW DID THEY DO THAT?

Wonders of the Far and Recent Past Explained

Caroline Sutton

A HILLTOWN BOOK

QUILL
New York

Copyright © 1984 by Hilltown Books

Permission to reproduce illustrations has kindly been granted as follows:

How did Rudolph Valentino spend his wedding nights?: The Museum of Modern Art/Film Stills Archive.

How did they know that the dodo was extinct?: Drawing by Lilly Langotsky.

How did they get the stones in place at Stonehenge?: Department of Environment, Great Britain.

How did James Joyce support himself while writing *Ulysses*?: Harley K. Croessman Collection of James Joyce, Morris Library, Southern Illinois University at Carbondale.

How did Dick Fosbury change the techniques of high jumping?: Wide World Photos.

How did they design the first car?: Drawing by Lilly Langotsky.

How did they pick the Four Hundred?: The Metropolitan Museum of Art; gift of R. Thornton Wilson and Orme Thornton, 1949. (49.4)

How did ballet dancers start dancing on their toes?: Courtesy of the Dance Collection, The New York Public Library at Lincoln Center, Astor, Lenox and Tilden Foundations.

How did Houdini escape from a packing case underwater?: Culver Pictures.

How did they build the Eiffel Tower?: Drawing by Lilly Langotsky.

How did they name the Edsel?: Courtesy of the Educational Affairs Department, Ford Motor Company.

How did they build Central Park?: Courtesy of the New York Historical Society, New York City.

How did Willie Sutton decide to rob banks? And how did he do it?: Courtesy of the Federal Bureau of Investigation.

How did the whooping crane learn to reproduce in captivity?: Courtesy of the International Crane Foundation.

How did they build the Great Pyramid at Giza?: Drawings by Lilly Langotsky.

How did Bach get the job of cantor at St. Thomas's Church in Leipzig?: Music Division; The New York Public Library at Lincoln Center; Astor, Lenox and Tilden Foundations.

How did they decide how tall to make the Empire State Building?: Photos by Louis Hine; courtesy of the Empire State Building.

How did they make Warren Harding the Republican presidential nominee in 1920?: Wide World Photos.

How did they design the Model T?: Courtesy of the Educational Affairs Department, Ford Motor Company.

How did men decide to wear neckties?: Trustees of the British Museum.

How did Beethoven compose when he was deaf?: Music Division; The New York Public Library at Lincoln Center; Astor, Lenox and Tilden Foundations.

How did Harpo learn to play the harp?: The Museum of Modern Art/Film Stills Archive.

How did they kill Marie Antoinette?: *Marie-Antoinette conduite au supplice*. Louvre Museum. Photographie Giraudon.

How did Hannibal cross the Alps?: Reprinted by permission of Thames and Hudson, Ltd., from *Hannibal: The Struggle for Power in the Mediterranean,* London, 1969.

All rights reserved. No part of this book may be reproduced or utilized in any form or by any means, electronic or mechanical, including photocopying, recording, or by any information storage or retrieval system, without permission in writing from the Publisher. Inquiries should be addressed to Permissions Department, William Morrow and Company, Inc., 1350 Avenue of the Americas, New York, N.Y. 10019.

It is the policy of William Morrow and Company, Inc., and its imprints and affiliates, recognizing the importance of preserving what has been written, to print the books we publish on acid-free paper, and we exert our best efforts to that end.

Library of Congress Cataloging in Publication Data

Sutton, Caroline.
How did they do that?

"A Hilltown book."
Bibliography: p.
Includes index.
1. Questions and answers. I. Title.
AG195.S87 1985 031'.02 85-6484
ISBN 0-688-05935-X (pbk.)

Printed in the United States of America

18 19 20

BOOK DESIGN BY BERNARD SCHLEIFER

ACKNOWLEDGMENTS

I WISH TO THANK Lilly Langotsky for her fine drawings, and Thaddeus Rutkowski for his contribution of articles on the discovery of Troy, the breaking of the Japanese secret code, and the making of the first accurate map of the world. I am grateful to all those who offered their advice and expertise, particularly David Kahn, author of *The Codebreakers*; Dr. Spyros Iakovidis and Dr. David Gilman Romano of the University Museum at the University of Pennsylvania; Zahi Hawass, chief inspector of the Giza Pyramids; Dr. Harry L. Shapiro of the American Museum of Natural History; Dr. David Knipe of the University of Wisconsin; Alexandra Munroe of Japan Society in New York; Richard Barringer of the Shipley School; Ryland Clark of the Collegiate School; David S. Staiger and Douglas A. Bakker of the Ford Motor Company; Joan Fordham of the International Crane Foundation; and Fred Oster of Vintage Instruments in Philadelphia. I would especially like to thank Brian Dumaine, Anderson Sutton, and James and Eloise Sutton, whose ideas and assistance were invaluable.

CONTENTS

HOW DID THEY DO THAT?

?

How did they spend $40 million making *Heaven's Gate?*

When Michael Cimino's 3-hour-and-39-minute epic was screened in New York in November 1980, United Artists' managers blanched, the critics brandished their pens, and Cimino hastened to explain that the film had been edited with undue speed to meet a release deadline. No doubt about it, his two-year obsession, his beloved vision of the 19th-century West, was a bomb—a shocking debacle for a hot director who'd managed to keep even the studio from seeing it until the last minute. The film is "something quite rare in movies these days," wrote Vincent Canby, "an unqualified disaster." And the inimitable Pauline Kael described it as a "woozy, morose mixture of visual virtuosity, overarching ambition, and slovenly writing," adding that "it's a movie you want to deface; you want to draw mustaches on it. . . ."

This monumental embarrassment to United Artists caused many film companies to take a hard look at the megabucks they were pouring into moviemaking with such enthusiasm. Other recent biggies included *Star Trek* at $40 million (one of the few that earned out and then some), *The Blues Brothers* at $36 million, and *Apocalypse Now* at $32 million. *Heaven's Gate*, it was estimated, would have to gross $140 million *just to break even*. The big studios suddenly wondered whether the time had come to curb their zealous directors, be they divinely inspired or merely self-indulgent.

The screenplay for Cimino's epic about a range war between cattlemen and immigrant farmers in the 1880s had been kicking around for about ten years. One studio president who rejected it called it a "a good yarn if you wanted to make a western" (which no one really did), while elsewhere it was criticized for its lack of humor, pace, and clarity. But then *The Deer Hunter* won Cimino the 1978 Oscar for best director and he became Hollywood's boy wonder overnight. The new management at United Artists decided to take on the project, which Cimino at one point claimed could be completed for under $8 million. The contract signed in the spring of 1979 allotted him $11.6 million.

That summer Michael Cimino, Kris Kristofferson, Christopher Walken, Isabelle Huppert, Jeff Bridges, producer Joann Carelli, and over 100 technicians and craftsmen packed up their belongings and headed for the sleepy town of Kalispell, Montana. From there the obsessive director launched his dream, with his mania for authenticity the sole beacon, the guiding force, the green light to spend—and keep on spending. The main street of Wallace, Idaho, was turned into boomtown Casper, Wyoming, of 1892; a fictional town of Sweetwater was erected in Glacier National Park; sets were built on the Blackfoot Indian reservation at the Canadian border and in the Flathead National Forest. More extravagant still was a 100-foot-long, 42-foot-wide skating rink built in Kalispell, complete with a huge woodburning stove that caused temperatures inside to soar to 100° and necessitated packing the cameras in ice. Two hundred fifty extras practiced on old-fashioned skates for six months—all for a skating sequence in the film of a few *minutes*.

"Every article of clothing, every structure, every sign," asserted Cimino, "is based on a photograph of the period." A 19th-century locomotive from Denver had to be rerouted all over the West because it was too big to go through most tunnels. Cimino himself combed this country and many others before finding a hat maker in Philadelphia who could make top hats to his satisfaction. He used 80 wagon teams and, as one crew member remarked, he "interviewed 300 horses for this movie." Twelve hundred actors and extras attended classes in bullwhipping, wagon driving, waltzing, and horseback riding, while one of France's most prominent young actresses, Isabelle Huppert, went to a bordello in

Idaho for three days to learn her role from the pros. Hoping to safeguard his brainchild from the label of western, Cimino added a brief epilogue on a yacht, which cost $300,000, and a lavish prologue featuring an 1870 Harvard graduation with hundreds of students waltzing on the lawn, brass bands, and boys in top hats rushing through the streets. (Ironically, Cimino's insistence on the real lapsed when it came to the plot, which deviates considerably from the historical facts.)

Cimino insisted that each scene be shot 20 or 30 times where most directors would settle for 4 or 5. Kris Kristofferson had to wield a bullwhip before the cameramen over 50 times. Cimino reportedly threw a party when he surpassed the film footage of *Apocalypse Now*, reaching an incredible 1.5 million feet. He spent 156 shooting days, thousands of man-hours, millions of dollars. The townspeople of Kalispell soon caught on that movie crews from Hollywood have bucks to spare, and theirs was a captive market. In negotiating for land use, for instance, producer Carelli told *American Film*: "After we made the deals, before production started, every landowner changed his mind. 'You can't use my land,' he'd say, 'unless you give me another $50,000.' What choice did we have? In some cases we did try to find new locations, and that would take more money, too. It was like holdup time without a gun."

If the powers that be at United Artists had a glimmering of the extent of Cimino's extravaganza, they hesitated to intervene. They were in too deep, and what would a subjective film grounded in a personal vision be without the one who envisioned it? So in 1980 the film had a brief stint before the public eye (which was stunned and amused), then was recalled, reedited, and rereleased in 1981, at which point critics still called it "murky" and "inhuman." With interest payments, advertising, and reediting costs on top of the initial estimated $36 million in production expenses, total costs hovered around $40 million, perhaps higher—a memorable catastrophe in the history of show biz.

?

How did they discover penicillin?

A visionary scientist made a serendipitous discovery and fought over a decade for the scientific community to accept it. When they did, the world received the lifesaving drug penicillin.

In 1928 bacteriologist Dr. Alexander Fleming, a modest man, son of a Scottish farmer, was working in his laboratory at St. Mary's Hospital in London, a quiet place resembling the back room of an old-fashioned drugstore. He was studying colonies of *Staphylococcus aureus*, a pus-forming bacterium, attempting to isolate a pure germ culture. One cool, damp summer day (precisely the conditions that molds take to) Dr. Fleming set aside several agar plates of his bacteria. When he later examined them under his microscope he detected mold spores, which had seemingly appeared out of nowhere, not only thriving there but dissolving his precious bacteria. "What had formerly been a well-grown colony was now a faint shadow of its former self," wrote Fleming in his notebook. Fleming might very well have thrown away the spoiled culture in irritation, for "when I saw the bacteria fading away," he later admitted, "I had no suspicion that I had a clue to the most powerful therapeutic substance yet found to defeat bacterial infections in the human body."

Nevertheless, Fleming decided to investigate. Removing a fleck of the mold, he placed it in a protein solution, where it grew rapidly. It is from this very culture that the molds producing the penicillin we use today are descended. The mold was a cluster of flask-shaped cells that resembled a brush, its spores forming at the tips of branching hyphae, or tubular filaments. Noting the structure of the mold cells, Fleming identified them as belonging to the *Penicillium* group, named from the Latin *penicillus*, meaning brush (they were later found to be *P. notatum*). He proceeded to implant the mold on plates with different pathogenic bacteria, some of which were affected dramatically (notably *gonococcus*, the cause of gonorrhea), others not at all (*B. typhosus, B. dysenteria*). When rabbits and mice were injected with a mold extract, their health and habits were not altered in the least. With mounting excitement, Fleming added penicillin to a sample of human blood serum. The white blood cells were neither injured nor restricted.

In retrospect, it is puzzling how resistant the world was to the drug it so desperately needed. News spread slowly and monies for production failed to appear. Many people were skeptical about the potential of the mold that had alighted by chance on Fleming's plate. Time and again the persistent scientist tried to interest his colleagues in penicillin, but doctors turned a deaf ear, disinclined to experiment on their patients with a common mold. Fleming made a detailed report to a group of bacteriologists, but they, too, failed to take interest in his find. Further problems resulted from the fragility of the mold; it was weak, impure, and easily destroyed by climatic and acidic changes. Huge amounts were necessary to obtain a concentration of penicillin sufficient for even one patient, and Fleming lacked adequate funds. Penicillin lay neglected for ten years, but Fleming did not allow his mold to die.

With World War II came an urgent need for antiseptics to battle infections of the wounded troops. Dr. Howard Walter Florey, professor of pathology at Oxford's Sir William Dunn School of Pathology, had heard of Fleming's mold and, against the severe odds, undertook further research. With a stellar team of 20 scientists and technicians, Florey recultivated Fleming's mold. For months this team tended huge vats of smelly, moldy broth, striving to extract the key ingredient. Dr. Ernst Boris Chain managed to dissolve out of the solution a brown powder, which was instantly lethal to some bacteria; actually the extract contained only about 5 percent penicillin in its pure chemical form. The scientists tested the substance on 80 different microbes. They determined that the blood fluids were not hostile to it and that white blood cells were not harmed or inactivated. They prepared a salt of penicillin (containing salts of sodium and calcium), which was more stable than the mold, and successfully cured mice injected with lethal doses of *Staphylococcus aureus*, *Streptococcus pyogenes*, and other bacteria. Their findings laid the foundation for penicillin therapy as practiced today.

In 1940 penicillin was used on the first human patient, a policeman with an advanced case of blood infection. For five days doctors administered the drug every two to three hours; penicillin passes readily out of the body in the urine and must therefore be replaced at regular, frequent intervals. The policeman had recovered significantly when the penicillin supply gave out and injections were halted. The infection flared up and ultimately took control. The British scientists could not possibly produce enough penicillin to be useful, months of work being required in order to save one life. In the second case, however, a young boy, also suffering from blood infection, received sufficient penicillin to recover.

With only five cases behind them, three of which had had successful results, the doctors were invited to America to carry on their research under the sponsorship of the Rockefeller Foundation. At the Northern Regional Research Laboratory in Peoria, Illinois, began another story: large-scale production of the invaluable drug.

?

How did they decide to make Washington the capital of the United States?

In June 1783, a horde of soldiers, still unpaid for their services in the War of Independence, stormed Philadelphia to voice their complaints to Congress. The country was then a loose union of 13 semi-independent sovereignties. There was no president, no money in the Treasury. The wise gentlemen in Philadelphia's Old City Hall decided it was high time to establish a federal city where angry soldiers might be handled in an organized manner, where the dignitaries of government might work efficiently and in peace.

For seven years the idea was bandied about. While many congressmen agreed on the need for a capital, they could in no way agree on its location. Northerners favored Philadelphia, Germantown (now part of Philadelphia), or New York. At one point a bill actually passed in both the House and the Senate designating Germantown, but when no speedy action was taken, the bitter debate was renewed. Southerners were more skeptical about the whole idea of a centralized government, fearing an oligarchy and particularly one that catered to the commercial interests of the North. If they had to have a capital, it had better be in the South. By this time there was a president, and *he* espoused a site on the Potomac.

Such was the impasse in Congress when yet another volatile issue took center stage, further threatening the union of the colonies. Alexander Hamilton, as Secretary of the Treasury, proposed that the federal government assume all state debts incurred in the Revolution. The northern states, whose debts were considerably higher than those in the South, thought it an excellent idea. Southerners, on the other hand, felt the federal government was wielding excessive power, interfering with duties that, by rights, were the responsibility of each state. They balked, and Hamilton's bill missed its mark by two votes. But the North would not accept the defeat and flatly refused to conduct business until it was reversed. Threats of secession were rampant.

A distraught Hamilton approached Virginia's Thomas Jefferson outside President Washington's house one day and walked him back and forth for half an hour, lamenting the chaos in the legislature, stressing the urgency of unity and the need to rally around the President. Jefferson, claiming to be a novice at such matters, decided to have Hamilton and a few friends for dinner the next evening and discuss a compromise over capons and wine. At this meeting, Jefferson agreed that the two votes should be rescinded to save the Union. "But," he later wrote, "it was observed that this pill would be peculiarly bitter to the Southern States and that some concomitant measure should be adapted to sweeten it a little to them." The tradeoff was the capital. ". . . It was thought," continued Jefferson, "by giving it to Philadelphia for ten years and to Georgetown permanently afterwards, this might, as an anodyne, calm in some degree the ferment which might be excited by the other measure alone. So two of the Potomac members (White and Lee, but White with a revulsion of stomach almost convulsive) agreed to change their votes and Hamilton undertook to carry the other point."

On July 16, 1790, the House passed an act by a vote of 32 to 29 establishing the seat of government according to Hamilton and Jefferson's scheme—it would stay ten years in Philadelphia "and then go to the Indian place with the long name, on the Potomac." This was Conococheague (or Conongocheague), named for a stream marking the site's northern boundary. The act, furthermore, empowered the President to set the exact boundaries of the federal territory, which was not to exceed ten square miles.

Why was George Washington so keen on the Potomac area? He envisioned the river's great commercial potential, reaching the tobacco market of Georgetown and points farther west via a proposed canal across the Cumberland Gap. As a boy he'd ridden in the surrounding hills and grown to love the lowlands of alder and the gentle, wooded slopes. He'd seen, too, that the Indians held council here, that *they* thought it was a fine place for legislative government.

The original district that Washington outlined included Georgetown to the north, Alexandria to the south, and the village of Hamburg lying beside the river's swamps, an area commonly called Foggy Bottom. Indeed, to many, the swampland wilderness seemed an unlikely milieu for the stately residences of government. Nevertheless, Washington clung to his vision of a magnificent city, planned by the reputable French military engineer Pierre Charles L'Enfant to rival Europe's finest. But when the Congressmen moved bag and baggage to Washington's "Emporium of the West" (as he called it), they were not quite so enraptured. In the early 19th century, more common appelations heard around town were "wilderness city," "Capital of Miserable Huts," and "A Mud-hole Equal to the Great Serbonian Bog."

?

How did they find the Dead Sea Scrolls?

In the spring of 1947 a Bedouin boy known as Muhammed the Wolf was herding his goats on the western shore of the Dead Sea near the desolate ravine the Arabs call Qumran. This may seem a strange place for the boy to have been, but he was among a group of contrabandists smuggling goats out of Transjordan to Palestine, detouring south to skirt the customs officers at the Jordan Bridge and to get water from the spring of Ain Feshkha. On this particular morning, one of Muhammed's goats strayed. The boy scrambled up a cliff after it, and there caught sight of a cave. Hesitant to explore it, Muhammed threw a stone into the dark opening and immediately heard the sound of something cracking, whereupon he fled. But curiosity nagged the goatherd, and he returned later, this time with a friend. The two boys entered the cave and found several tall clay jars and fragments of others. On lifting off the bowl-shaped lids, they were struck by a terrible smell. Inside were mysterious packages, dark and oblong. These they took out into the blazing sunlight and opened. First there was a layer of black wax or pitch, and beneath it strips of linen. Unrolling the cloth, the boys found long manuscripts, made up of thin sheets sewn together. Although the scrolls were faded and crumbled in spots, the parallel columns were in general very clear. The strange script was not Arabic, however, so the boys had no inkling of the nature of their find, but figured it might bring something on the black market in Bethlehem.

The first merchant the boys approached refused to pay their asking price of £20. The second was a Syrian, who suspected that the language was ancient Syriac, and sent word to the Syrian Metropolitan at the Monastery of St. Mark in Old Jerusalem. This dignified gentleman, Mar Athanasius Yeshue Samuel, with his large black beard, voluminous black robes, and onion-shaped satin miter, expressed immediate interest in the find from the isolated region where no one had lived since the early Christian centuries. Upon examining a sample, the Metropolitan determined the manuscripts were leather or parchment, and the script Hebrew. He sent word that he was eager to buy the entire lot, but the contrabandists were already on the road again. They returned in several weeks and the Metropolitan arranged for the scrolls to be delivered. All morning he waited. Finally despairing, he went to lunch. It was then, of course, that the Bedouins appeared. The priest who greeted them noted their shabby appearance, doubted their claims, and turned them away. On hearing this, the Metropolitan was irate. He kept on the trail of the Bedouins and learned that a Jewish merchant had arranged to buy the manuscripts and had invited the boys to his office in the New City, which was largely Jewish.

At this time the Arabs and Jews were in conflict in Jerusalem. The city was divided and the Jewish sections were under martial law. The clever Syrian who had arranged the sale to the Metropolitan thus easily convinced the Bedouins that this Jewish merchant was out to trap them, that he would rob them and throw them in jail. The fearful Bedouins left five scrolls with the Syrian right there and then. Later they took them, along with other fragments, to the Metropolitan, who bought them for a reported £50.

The Metropolitan proceeded to send a merchant and a priest out to the cave to verify the Bedouins' story. The two men spent an uncomfortable night in the stifling August heat, found the pottery fragments and linen, but soon retreated, bringing no samples of their finds. The Bedouins, it was discovered later, had been using two of the ancient jars to carry water.

At this point the Metropolitan Samuel launched a long search for the identity of his scrolls, and in the face of uninterest, discouragement, and ridicule, he maintained an absolute faith in their ancient origins. He approached a Syrian in the Palestinian Department of Antiquities and a French priest at the Dominican École Biblique, a center of archaeological research in Old Jerusalem, but neither expressed much interest. The two most prominent archaeologists in the region, G. Lankester Harding of the Department of Antiquities of Transjordan and Père Roland de Vaux of the École Biblique, proved inaccessible. Others at these institutions insisted that the manuscripts could not possibly be 2,000 years old. Nor did they pursue the matter even after a Dutch scholar, Father J. van der Ploeg, read one scroll and recognized it as the book of Isaiah. When the professor of Hebrew at the American University in Beirut was found to be away, the persistent Metropolitan got his own Hebrew dictionary and attempted to read the text himself. At this time, a Hebrew scholar, Tovia Wechsler, had a look at the manuscripts and decided they dated only from 1929, that they had been stolen from a Palestinian synagogue during Arab riots against the Jews.

In November 1947, the head archaeologist of the Hebrew University, E. L. Sukenik, returned to Jerusalem and learned that a dealer in Bethlehem had some manuscripts from a cave near the Dead Sea. This dealer was the first merchant the Bedouins had approached; he had learned later that the scrolls might be valuable and had managed to get hold of three. Throughout late November and December, Sukenik visited the dealer and examined the scrolls. "The script seems ancient to me . . . ," he wrote in his diary on November 27. "Is it possible?" That evening news of the partition of Palestine was announced. The situation was extremely tense, but Sukenik nevertheless made the treacherous journey to Bethlehem and brought back three scrolls and some fragments. On December 21, he obtained another. Amid the shelling of the offices of the Jewish Agency in New Jerusalem, Sukenik held a press conference there at which he announced the magnitude of the discovery. He held these scrolls to be the first ancient Hebrew manuscripts ever revealed, and he dated them to the 1st or 2nd century B.C.

The Metropolitan Samuel, meanwhile, was still searching. In February 1948 he took his scrolls to the American Schools of Oriental Research and showed them to Dr. John C. Trevor, who was acting director. Trevor could not make an immediate estimate of their age, but he did compare the texts with other samples of early Hebrew script. "One glimpse at the picture of the British Museum Codex from the ninth century," recalled Trevor in the *Biblical Archaeologist*, "assured me that these scrolls were far older. The next slide was of the Nash Papyrus, a small fragment in the University Library at Cambridge containing the *Shema* and the Ten Commandments. . . . The similarity of the script in the papyrus and the scrolls was striking." This Nash Papyrus was generally believed to be the oldest extant Hebrew manuscript, dating from somewhere between the 2nd century B.C. and the end of the 1st century A.D. The exhilarated Dr. Trevor, the Metropolitan, and other Syrians anxiously waited out a power failure and eventually managed to photograph a good part of the scrolls. Prints of the Isaiah scroll were sent to Dr. W. F. Albright of Johns Hopkins, an outstanding archaeologist and expert on the Nash Papyrus. By return mail, Dr. Albright sent the confirmation that the Metropolitan had long awaited and expected: "My heartiest congratulations on the greatest manuscript discovery of modern times! There is no doubt in my mind that the script is more archaic than that of the Nash Papyrus. . . . I should prefer a date around 100 B.C. . . . What an absolutely incredible find! And there can happily not be the slightest doubt in the world about the genuineness of the manuscript."

The aftermath of this discovery was equally extraordinary. Once the war was over, Père Roland de Vaux and G. Lankester Harding undertook a thorough search of the cave in which the scrolls had been found. In February 1949 they turned up fragments of a Roman lamp and a Roman cooking pot, as well as Greek pottery, which indicated the scrolls must be at least as old as the 1st century A.D. The mass of shards suggested that the cave might once have held some 200 scrolls. Three years later, the two archaeologists returned to the Dead Sea along with the Bethlehem chief of police and some Bedouins, who led them to four large caves, 15 miles south of the first. Bedouins came scrambling out of these caves, and the team found traces within of human habitation from the 4th millennium B.C. Objects of the Bronze and Iron ages were found alongside many Roman artifacts—lamps, combs, buttons, spoons, bowls, and coins. There were fragmentary manuscripts in Greek, Latin, Hebrew, and Aramaic. Père de Vaux concluded that the caves had once been a base of Jewish resistance against the Romans.

There seemed little connection with the first discovery, however, so Harding and De Vaux moved nearer to Qumran and explored the myriad caves in its vicinity. Out of 267 caves, they discovered pottery and vestiges of human habitation in 37. And in 25 of these, the pottery found was precisely the same as that in which the ancient sacred writings had been stored. Scrolls were found buried in dirt, but thus exposed to the elements, most had disintegrated badly. Thousands of pieces were collected, and the world grew increasingly excited. For here, it became clear, an ancient people had hidden not one isolated document, but an entire library, including nearly all the books in the Bible, several apocryphal works, and literary writings—all of which might still lie concealed from the modern world had it not been for one wayward goat.

?

How did Rudolph Valentino spend his wedding nights?

Women everywhere melted at the sight of the inimitable Valentino in his silent-film role of the seductive Sheik. But although thousands yearned for his affections, his bride was apparently not among them. Jean Acker and Rudolph Valentino, young and new to Hollywood, married on a sudden impulse in 1919 and realized the same day that their vows were not to be. When it came to the ritual crossing of the threshold, Jean darted ahead and slammed and bolted the door, leaving her hapless husband out in the cold. She eventually sued him for desertion, oddly enough, and he countersued. When it became known that the marriage had never been consummated, an interlocutory decree of divorce was issued.

A designing Valentino eyes Agnes Ayres in The Sheik, *1938.*

The first hours of Valentino's second marriage were no improvement on the first. Captivated by the love of his life, Natacha Rambova (whose real name was Winifred Shaughnessy), he plunged into marriage with her before his divorce was final. Immediately charged with bigamy, Valentino landed in jail for a few hours. Bond was posted and "the Great Lover" had to assure the court that, once again, his marriage had not been consummated. The two had to live apart in the ensuing months and remarry only when the divorce was completed, giving Valentino yet another shot at living up to his mythical image on his wedding night.

?

How did they shoot up Faye Dunaway in *Bonnie and Clyde?*

When the police ambush the infamous duo in the climactic scene of *Bonnie and Clyde* (1967), Faye Dunaway writhes and twists beneath a deluge of fire in a slow-motion death that looks agonizingly real. While the motions are all her own—special effects don't always preclude the necessity of good acting—some technical tricks did enhance the image.

Both the car and Faye Dunaway sitting in the front seat were plugged by a shocking cataclysm of bullets, and both were wired with scores of little explosives to simulate the impact. Imagine getting dressed for a routine day's work in clothes rigged with electric charges, and only little metal plates serving as shields between them and you. That's what Faye Dunaway had to do, day after day, changing from one set of rigged clothes to the next for the different takes. Where each explosive was to rip through her clothes, the fabric was thinned with sandpaper to give the blast maximum effect. "She looked like a telephone switchboard with all those wires coming out of her," recalls special-effects artist Danny Lee, who orchestrated the scene. Holes were punched in the car and these were filled with similar charges, embedded in putty. When the police opened fire, the charges were set off successively, leaving the holes both in the car and in Dunaway's clothes as she jerked and convulsed under the simulated impact. In later films like *The Godfather*, plastic capsules or rubber pouches of synthetic blood were added to each metal plate.

?

How did they know that the dodo was extinct?

No one has seen a dodo for three centuries. Furthermore, no one *ever* saw one anywhere except on the remote volcanic island of Mauritius in the Indian Ocean.

When the Portuguese arrived on Mauritius in 1507, bountiful game was there for the plucking, and so they plucked. The dodo

The ill-fated dodo.

was among the island's many species that had lived until then in a paradise without predators. Specialization had brought it to the point of decadence, for the birds were obese, weighing some 50 pounds each, with paltry little wings and a stubby tail. Little did this poor bird know about the arts of defense, and the Portuguese called it *doudo*, meaning simpleton or fool. Not only did the new settlers and ensuing Dutch, French, and British sailors dine frequently on the dodo, they let loose pigs, dogs, monkeys, cats, and rats, which wreaked havoc on the indigenous species. The pigs tramped and snuffled through the underbrush, destroying precious habitats of ground-nesting birds. The monkeys easily found the dodo's eggs—each female apparently laid only a single egg each season, exposed to the world in all its glory on a carpet of grass in the forest.

One hundred seventy-four years of colonization thus put a gruesome, irrevocable end to the species. The last sighting was in 1681. The dodo became the first species for whose extinction man was clearly culpable. Today the dodo is represented by only a head and foot, preserved in the Ashmolean Museum at Oxford University, and a quantity of bones, housed in the British Museum, Paris, Leyden, Brussels, Darmstadt, Berlin, and New York.

Since the 17th century, 22 of Mauritius's 33 species of indigenous birds have followed the plight of the dodo. Today ornithologists are desperately trying to guard and protect those which remain. The luminous green echo parakeet with its brilliant red beak numbers only five. Wild pink pigeons are dangerously few, and an elegant Mauritius kestrel winging over the island's craggy volcanic crater may be one of only a dozen of its kind in the world.

?

How did they put out the San Francisco fire of 1906?

The terrible tragedy is they didn't. The fire raged for three full days, devouring all in its three-mile path, razing the city, leaving hundreds of thousands homeless and helpless.

At 5:13 A.M. on April 18, 1906, the San Andreas Fault shifted and heaved with enough force to buckle the streets of San Francisco, shatter her mains and sewers, fell power lines, and turn buildings to rubble. Two lesser earthquake shocks followed the first. By then the city's electrical wires had touched off an inferno. People fled their beds and rushed into the streets, screaming, crying, or gazing about, stunned, wondering if they could still be dreaming. All electricity was cut off; Western Union shut down; all transportation halted. The streets reeked from ruptured sewer pipes, and broken mains rendered the fire department helpless. Rich and poor alike streamed through the streets beneath a cascade of bricks, cornices, and corners of buildings. The $7-million city hall collapsed; the new post office was shattered; the entire south side of Market Street from Ninth Street to the bay was ablaze. Here stood the city's prized Grant, Parrott, Flood, Examiner, and Monadnock buildings, along with the majestic Palace and Grand hotels. Commercial buildings north of Market Street also succumbed, and at the Opera House the lavish scenery and costumes of the visiting New York Metropolitan Opera fueled the flames like newspaper in a campfire. There followed a deafening blast as the gas house exploded and at this, reported *The New York Times*, "a feeling of despair overcame the men who were performing the rescue work."

Although most of the city's fire alarms had failed, the corps of 600 fire fighters had gone to work immediately. But the extent of the fires, the speed with which they spread, and the shortage of water everywhere except along the bay front were insurmountable obstacles. Firemen resorted to dynamite to create firebreaks, and for several days interminable blasting added to the mayhem.

But the tumbled buildings only caught more easily, the fire lashing through the kindling and streaming out of windows.

Infantry, cavalry, and police, several thousand strong, patrolled the streets under strict orders from the governor and commanding general to shoot down vandals on sight. Thirty-five were shot and killed trying to loot the ruined United States Sub-Treasury on Commercial Street. Chief of Police Doonan ordered all saloons closed to prevent drunks from becoming unwitting targets. The entire city was placed under martial law.

On April 20 *The New York Times* headline read: BOMBARDMENT A MILE LONG FAILS TO SAVE SAN FRANCISCO. MANSIONS WRECKED BY CANNON IN LAST STAND ON NOB HILL. An artillery stand against a natural disaster seems peculiar and even pathetic, but that was the only course the forces in San Francisco could pursue. While thousands huddled in the parks, hungry and terrified, the fires roared and shells boomed. Cannon fire destroyed the beautiful homes east of Van Ness Avenue and a mile of buildings along the avenue, creating a 500-foot-wide swath that was generally successful in stopping the fire, although flames leaped across at certain points. Meanwhile, the gusty winds kept shifting and the fire reached the luxurious abodes of Nob Hill. At the "fireproof" Hotel Fairmont, sailors assisted the firemen and made citizens work at gunpoint. Finally, the last remaining barrels of gunpowder in the area were called in from the Presidio, Fort McDowell, and Alcatraz. The millionaires from California's Gold Rush bonanza trickled out on the streets to watch the cannons take aim at their mansions. These palaces with all their furnishings were shot to ruins.

Eventually the fires in the Nob Hill and Telegraph Hill areas and the Mission district turned back on themselves and burned out. Little remained except the skeletons of some new steel buildings, a portent for the future. The 21st of April brought rain to the smoldering ashes of 25,000 buildings. Nearly five square miles of the city had been destroyed in three days, with property damage running about $400 million. Total loss of life was never confirmed; estimates range from 300 to 600. The survivors faced the overwhelming challenge of rebuilding, which they met with commendable spirit and success.

?

How did everybody start twisting?

Chubby Checker may have popularized it, but he did not invent it. The twist step goes way back to the early black minstrel and medicine shows. At the turn of the century, Jelly Roll Morton used to sing down in New Orleans about "sis . . . out on the levee doin' the double Twis'." And pianist Perry Bradford wrote a song in the late '30s called "Messin' Around" that tells you to put your hands on your hips, lean back, and "twist around with all your might./Messin' round, they call that messin' round."

Although white teenagers and toddlers, weight watchers and debutantes caught the craze in the '60s, the dance was African or Afro-American in style. Obviously, the buttocks were a focal point, and, as in the shimmy and the black bottom, the dancer bent his legs, bringing him closer to the earth, rather than gliding about erect and poised as in many formal European dances.

Black ghetto kids were twisting in the '50s as part of a group dance called the Madison. At 19, Chubby Checker picked it up and toured the country. He sang "The Twist" (which was first recorded in 1959 by Hank Ballard) and shimmied across the stage, twisting high, twisting low, and showing white Americans how to "get down."

Most of them didn't budge, though, till the Jet Set had sanctioned this peculiar dance in which you didn't even touch your partner. (Perfume companies were despairing over this.) Somehow New York's trendiest discovered the Peppermint Lounge, a sleazy hole-in-the-wall in Times Square frequented by hoods and hookers. There, singer Joey Dee and the Starlighters let loose with the twist and the place went wild. "It took me an hour to get a drink," one customer complained; "even the waitresses were twisting." The press, including gossip columnist Igor Cassini, more widely known as Cholly Knickerbocker, discovered the Peppermint Lounge. Celebrities soon displaced the original clientele, and five mean bouncers manned the doors. From this seedy spot the twist spread like radar in every direction—to school gymnasiums and small-town bars, to Paris nightclubs and, allegedly, to the White House.

It was a dance revolution and a sexual revolution, of which not everyone approved. Roseland ballroom would have none of the graceless "fad," and dancer and painter Geoffrey Holder called the twist "synthetic sex turned into a sick spectator sport." But the overwhelming majority were ready to leave the lindy and the waltz in the dust. Except, that is, in Russia:

"The Twist is characteristic of pleasure-sated people in the free world who, to our high-minded cultural society, resemble patrician practitioners of orgies in ancient Rome." This was the official report of the New York correspondent for Moscow's *Literary Gazette*. "It is a dance performed to a noisy variation of rock 'n' roll played by drums of the stone age."

?

How did they get the stones in place at Stonehenge?

The massive stones on England's Salisbury Plain have stood for over 4,000 years, witness to the wanderings of prehistoric nomads, to the rise and fall of tribes, to rites, mysteries, and ceremonies forever eclipsed. Well before the complexity of its design was revealed, the elegance and imposing size of Stonehenge inspired scores of theories about its ancient origins and purpose. Was it a defensive structure? A cemetery? A druidic temple of the moon? A ring of Irish petrified giants, magically carried from abroad, as one imaginative legend suggests?

As early as 1740 Dr. William Stukeley detected the alignment of the main axis with the midsummer sunrise. Since the late 19th century, an increasing number of astronomical links have been established whose sophistication and precision contradict the widespread notion of a prehistoric race more apelike than human. Astronomer Gerald S. Hawkins, a Stonehenge authority and author of several books on the subject, perceived in the arrangement of stones an elaborate scheme of moon and sun alignments: "Stonehenge I [the earliest construction] had 11 key positions," he wrote in *Stonehenge Decoded*, "every one of which paired with another, often more than one other, to point 16 times to ten of the twelve extremes of the sun or moon; Stonehenge III [built later] with its five trilithons and heel stone axis pointed 8 times to eight of those same extremes." Hawkins further suggested that the monument, so subtle in form, so ingenious in function, acted as a computer to predict the terrifying natural phenomena of solar and lunar eclipses.

The why of Stonehenge may never be unveiled, but archaeological discovery of tools, examination of materials, and meticulous study by authorities Richard J. C. Atkinson and Gerald Hawkins provide more tangible evidence of the how. The enormous task of erecting Stonehenge was carried out by different prehistoric peoples in three major construction stages, ranging over a period of at least 500 years.

The period of first building, Stonehenge I, began around 2200 B.C., as established by radiocarbon dating techniques. At this time, Late Stone Age (Secondary Neolithic) people, probably hunters and farmers from the Continent, built a nearly perfect circular bank 380 feet in diameter. Within it they dug, with pickaxes of red deer antler and shovels of oxen shoulder blades, a roughly circular ditch, originally a series of separate pits 10 to 20 feet wide and 4½ to 7 feet deep, now thought to be quarries. Inside the ditch they piled an impressive circular bank of hard white chalk, 6 feet high, 20 feet wide. The two banks and ditch were left open to the northeast and a huge 35-ton "heel stone" was placed on the entranceway 100 feet outside the enclosure. The dramatic appearance of the midsummer sun over this stone must have inspired celebration and enhanced the power of the priests. To this day it excites visiting crowds. John Aubrey in the 17th century was the first to spot a sequence of holes within the enclosure, two to four feet deep, deliberately refilled with chalk rubble, bone pins, and cremated human bones. These 56 Aubrey holes, whose purpose is yet unclear, were meticulously placed along a circle 288 feet in diameter. At four of the holes there rose enormous "station stones," two of which remain. The longer sides of the rectangle suggested by these stones were precisely perpendicular to the summer sunrise line, and the diagonals intersected at the center of the circle.

In approximately 1700 B.C., the Beaker people added to the

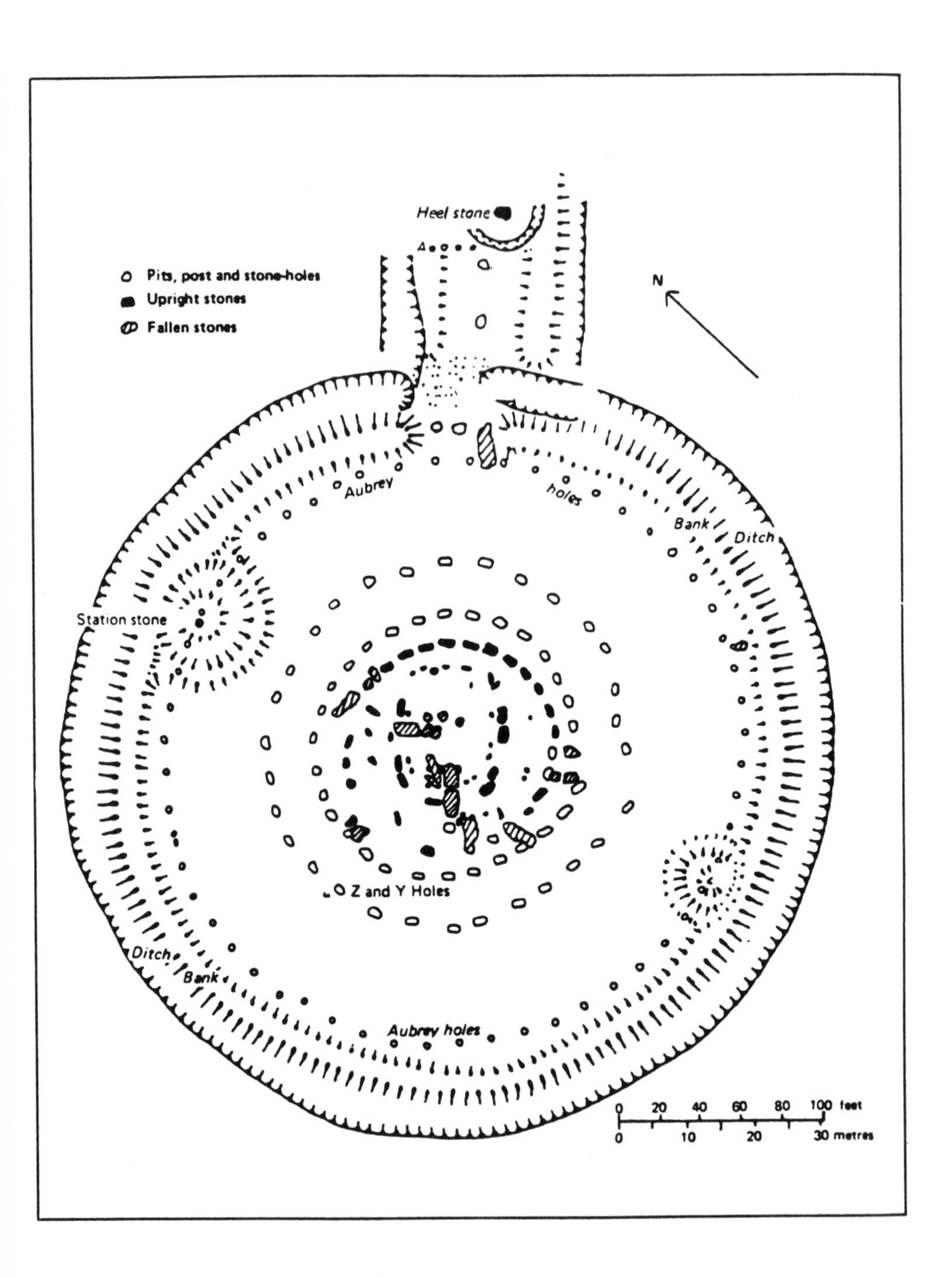

Plan of Stonehenge.

extraordinary work of their predecessors. Within the enclosure they erected at least 82 bluestones (made of an igneous rock with a bluish tint), weighing up to five tons each, in two concentric circles, the stones of the inner circle perfectly aligned with those of the outer. At this time, too, a majestic 40-foot-wide avenue was formed northeast from the monument and eventually branching eastward to the Avon River 2 miles away.

Approximately 1,000 years later, the powerful, rich, and sophisticated people of Wessex took down the double circle of bluestones and replaced them with a different pattern of far more massive sarsen boulders of natural sandstone. At the center they erected a horseshoe of five trilithons (from the Greek words meaning three stones), each comprising two uprights and a lintel, or crosspiece. Here stand the largest of the stones brought to the monument; one sarsen upright measures 29 feet, 8 inches, and weighs 50 tons—the biggest prehistoric stone fashioned by hand in England, according to Hawkins. Around the horseshoe, the Wessex people built a ring of 30 sarsen uprights with connecting lintels, again left open to the northeast. The 25-ton uprights were uniformly placed, says Hawkins, with "an average error of less than 4 inches." The tops of the uprights were scooped out to hold the seven-ton lintels, which were further secured by a mortise and tenon system: a knob nine inches high was carved on the top of each upright to fit snugly into a hole dug on the bottom of the lintel, one hole near each end—a staggering feat when one considers the precision of measurement and placement of the stones, and the skillful carving achieved with the roughest stone tools. A pale green boulder of micaceous sandstone, 16 feet long and 3½ feet wide, was placed before the central trilithon. This has been named, somewhat misleadingly, the altar stone; neither its original position nor its purpose is yet fully explained. Somewhat later, the people erected a bluestone horseshoe of 19 uprights within the sarsen horseshoe, and a rough circle of bluestones (numbering 59 to 61 according to Atkinson) between the sarsen horseshoe and the sarsen circle. These may originally have been intended for the corresponding number of holes (called Y and Z holes) lying well outside the enclosure, in which pottery shards, chips of bluestone, and fine soil have been found. Finally, all activity at Stonehenge stopped, and the site was probably abandoned around 1400 B.C.

The story of how they got the stones in place must start (now that one has some notion of the meaning of "in place") with how they got the stones to begin with. The boulders were not just lying around like so many building blocks on the empty plain. Thousands of people had to *haul* them from somewhere—without wagons, trucks, cranes—and this, in itself, is an astonishing story.

The bluestones at Stonehenge comprise five different types of stone. Dr. Herbert Thomas of the British Geological Survey established in 1923 that in only one region do the three most dominent types occur together naturally, the other two appearing nearby. This region is the Prescelly Mountains of Wales, 130 miles from Stonehenge as the crow flies, *240 miles* by the sea and land route probably taken by those early builders. This route went southwest to Milford Haven (where the mica sandstone of the altar stone is found), along the coast in the Bristol Channel, up the Avon and Frome rivers, overland to the Wylye River, down this waterway to Amesbury, and along an avenue to Stonehenge—25 land miles, 215 water miles. The latter, by far the easiest part, was probably accomplished with a number of rafts or simple canoes, joined with crosspieces, poled along by several men. In 1954 the BBC broadcast a reenactment of the transporting of the stones, using concrete replicas. The experiment determined that 24 men were required to haul a 3,500-pound load up a 4° slope—16 men per ton—using a system of sledges and rollers, the same method used in the construction of the Egyptian pyramids. While one team strained on ropes of twisted hide attached to the sledges, others had to continually advance the rollers from back to front, and also clear the way.

If hauling the bluestones seems arduous, imagine the task of dragging 80 sarsen stones, averaging 30 tons each, 20 miles from the Marlborough Downs north of Stonehenge. Here the sandstone boulders are plentiful and little quarrying was required. The stones were sized roughly, then cut either by sharp wedges inserted into cracks or by hot and cold stresses applied along a break line, followed by bashing with 60-pound stones. Then the boulders were mounted on sledges. Using the formula of 16 men per ton, Hawkins estimates that 800 men were needed to haul the sarsens, with 200 more on hand to move the rollers—that a total of 1,000 haulers worked seven full years to accomplish the task.

At the Stonehenge site the builders dressed and polished the stones using mauls and other rough stone tools. Atkinson estimates that the 3 million cubic inches of stone removed from the sarsens, (and there may have been more) through bashing and grinding occupied 1 million man-hours of labor.

To erect the unwieldy stones, a hole was dug, the depth depending on the size of the stone. (In some cases, the aboveground portion of the stones had to be uniform.) An angle of 45° was formed on one side of the hole, thus creating a ramp along which to slide the stone. Wooden stakes were placed in the hole opposite the ramp to protect that side from the end of the stone. A team of 200 men rolled the stone over the hole, tipped it down the ramp, then heaved it up using ropes attached to the upper end of the stone and run through a high wooden scaffolding opposite the ramp. Or, at this point, the stone may have been levered into place by gradually building up a fulcrum of stones and wood. Once it was in place, the surrounding hole was rapidly filled and the stone allowed to settle for some months. The placement of lintels was more awesome still. A massive ramp of earth may have been engineered and the lintels pushed and pulled and dragged with ropes up the 40° incline to the top of the uprights. But since no archaeological evidence exists for these mounds, a more likely method was the use of a latticework made of logs. A lintel was positioned in front of two uprights and two logs placed perpendicular to it. The rock was then rolled onto the logs, and an extension of the logs was added where the stone had been. On top of this the workers laid two more layers of logs, the first parallel and the second perpendicular to the lintel. The huge stone was rolled to this higher platform and another, still higher, was constructed. After rolling the lintel countless times, the lattice tower was as high as the uprights and the lintel could be properly placed, the holes fitting over the knobs.

In all, Hawkins estimates, a total of 1.5 million man-days of labor were necessary to erect the beautiful and mysterious monument.

?

How did Marie Antoinette "let them eat . . ." croissants?

When Marie Antoinette was wallowing in riches in her Petit Trianon and her husband, Louis XVI, was gorging himself regularly on eight-course meals, many in Paris were hungry and some were starving. They found even bread, the staple of the French diet, beyond their means. Hearing of this, the frivolous queen is alleged to have retorted, "Why, then, let them eat cake."

There is, in fact, no proof that Marie Antoinette ever said this. The remark was heard in France many years before the French Revolution. The Comte de Provence humorously attributed it to his great-great-grandmother, wife of Louis XIV, while others credited Madame Sophie, daughter of Louis XV. Some sources maintain that Jean Jacques Rousseau was the originator of the famous phrase in his *Confessions*, but there it is spoken by an anonymous princess, who advises the populace to eat brioche. Rousseau's princess said this in 1740 and his book was published in 1766—when Marie Antoinette, 11, was playing in her mother's palace in Austria.

Which brings us to another little-known fact, which helps ameliorate the infamous reputation of the Queen. If she brought the French misery and oppression, she also brought the croissant. This delicious, flaky roll, now associated with the *haute cuisine* of France, originated with the bakers of Vienna in 1683. The Turks were storming the city at about this time and some say the bakers made *kipfel*, or crescent-shaped rolls, in obeisance to the invaders, whose flag displayed a crescent moon. But Viennese bakers today recall the situation differently. Their predecessors, they insist, working late one night in 1683 heard the Turks tunneling under the city and alerted the Austrian guards, who abruptly halted the siege. The Holy Roman Emperor Leopold I asked the courageous bakers to make *kipfel* in celebration of Vienna's victory. They were still doing so a century later when the young princess Marie Antoinette introduced the buttery crescent, or *croissant*, pastry to the French.

?

How did they block Roosevelt's plan to pack the Supreme Court?

Although Franklin Roosevelt was reelected to the presidency in 1936 by an overwhelming majority (he carried every state except Maine and Vermont), there was one very significant place where the majority opposed him: the Supreme Court. Social ills plagued the country and Roosevelt was determined to extend his New Deal, but the judicial veto threatened many of his programs. To date, the Court had found New Deal legislation unconstitutional in seven of the nine cases brought before it. The vote was close, however: 5–4 in three of the cases, 6–3 in two others. The minority that stood by the President and Congress comprised the most distinguished members: Louis D. Brandeis, Benjamin N. Cardozo, Harlan F. Stone, and occasionally Chief Justice Charles Evans Hughes.

"Courts are not the only agency of government that must be assumed to have the capacity to govern," said Stone to the conservatives, and he stressed the Court's need for self-restraint, since this was the only check on its powers. The President echoed his sentiments, claiming that the Court had "cast doubts on the ability of the elected Congress to protect us against catastrophe by meeting squarely our modern social and economic conditions." He feared for the survival of several pieces of legislation integral to his program: the Wagner Act, giving new authority to the National Labor Relations Board; the Holding Company Act, which blocked financial concentration in public utilities; the Social Security Act, establishing a national plan of old-age and survivors' insurance; and a federal-state plan of unemployment compensation. Roosevelt complained of the Court's narrow reading of the Constitution and set out "to save the Constitution from the Court, and the Court from itself."

Roosevelt and Attorney General Homer Cummings concluded that the obstacle of the Court majority would have to be surmounted, one way or another. They considered a constitutional amendment, but quickly abandoned this in view of the time such a process would consume. They finally decided—outrageous as it would appear to many Americans—to tamper with the actual makeup of the Court. Under the rather hollow argument that the Court was suffering from too great a burden of work, Roosevelt suggested the appointment of another justice for every justice aged 70 or over. Chief Justice Hughes quickly proved that the President's overwork argument was specious. Roosevelt then took a more straightforward and ingenuous approach, but his plan and slick attempt at executing it had already alienated many former supporters and scandalized his opponents. Democrats in Congress stood up against the President, and the nation was divided. A prolonged fight might well have ensued, embarrassing the President and halting once and for all his much needed social reforms.

Fortunately for all concerned, the Court literally did save itself from itself in an unexpected reversal of opinion toward the New Deal legislation. In the spring of 1937 the Court endorsed a Washington State minimum-wage law and upheld the Wagner Act. Clearly, the bastion of conservatism was weakening, and Roosevelt saw no need for additional members—particularly when one conservative justice resigned, giving the President an opportunity for a new appointment. Although the attempt to pack the Supreme Court was ultimately defeated, Roosevelt maintained that he had only lost a battle—he had won the war.

?

How did the Indians decide that cows were sacred?

The sight of cows, however emaciated, lumbering through the sultry and congested streets of Calcutta, is a peculiar one to the western eye, and many deride the ban on killing cattle in a country where millions go hungry. But what appears an arbitrary and irrational practice arose not only from religious beliefs, but from the pragmatic necessity of protecting a vital source of food.

In the 2nd millennium B.C. the Aryans came to India, bringing with them their own cattle, but because of the harsh climate these soon were replaced by the native breed. Dairy products became very popular with the peoples of India, and the Aryans recognized the need to protect their limited source. The text of the Rig-Veda, which dates from the Aryans' first period in India, designates all goats, horses, sheep, and buffalo as proper food, but only *barren* cows. The milk, butter, and cheese gained from a single cow could sustain far more people for a longer period of time than could its meat.

This practical regard for the cow added to an already existing, very ancient religious esteem. South Asian artifacts that predate the Indus Valley cities (c. 2500–1500 B.C.) by some millennia display a religious veneration of the humped bull. So the later waves of Indo-Iranian speakers were adding *another* dimension to an indigenous symbol.

In 1000 B.C. a strict religious code, recorded in the Atharva-Veda, prohibited all eating of meat. Barren cows were then turned over to mendicant Brahmins, the Aryan priests. Over a period of 300 years this stricture was relaxed and gradually it became the custom to kill cattle to sacrifice to the gods, or to meet the demands of hospitality. The Brahmin priests of 700 B.C. began abusing this freedom, however, sapping the country's wealth by demanding ever more cattle for ritual sacrifice. There arose in opposition two new religious leaders: the Buddha and Mahavira. Both denounced violence, the Aryan caste system, and the slaughter of animals. The followers of Buddha would not permit animals to be killed for them. The Jains, or disciples of Mahavira, were far more stringent, scrutinizing even fruits and vegetables before eating them to avoid killing insects. Their fanaticism stemmed not from love of the fruit fly, however, but from a belief in transmigration of the soul, whereby even an insect might incorporate a human soul—though undoubtedly one that had not behaved too admirably in a former life.

By 100 B.C. the influences of Buddhism and Jainism had made their mark. The Brahmin priests of the Vedic faith were now performing rituals without animal sacrifice. The old ban on killing cows was resurrected. Vegetarianism increased throughout the country, particularly in the south under the control of the orthodox priests. Some communities continued to eat goat and chicken, but the cow was protected. Scarcity of supply complemented the religious code. Even in the cities the wealthy refrained from eating beef; milk and yogurt were important ingredients of Indian cooking, and the greenish milk of the buffalo offered a poor substitute.

The inviolability of the cow was emphasized during the territorial expansion of the 5th century A.D., when settlers had little but a few cattle and rudimentary agricultural knowledge. Their Brahmin advisers strictly forbade killing the cows.

Hindus believe the cow, as provider of milk, symbolizes motherhood and that this is the basis and origin of its veneration. An ancient legend tells that the cow Surabhi, mother of all cows, was one of the treasures churned from the cosmic ocean. The "five products of the cow"—milk, curd, butter, urine, and dung—all embody great purifying potency and must be protected and sustained. (All five have practical value as well: dung is burned for fuel, particularly where wood is limited, and urine is occasionally used for its medicinal and cleansing properties.) Since no goddess represents the cow, the live animal itself is revered. Slaughter is equivalent to deicide.

Several incidents in more recent times point to the seriousness with which many Indians maintain their reverence for the cow. In 1857 the British introduced the Enfield rifle, whose cartridges had to be bitten open before loading. These cartridges were covered with grease and at one point rumors burned through the ranks of Indian sepoys that the grease was either beef or pork fat. This outraged both Hindus and Muslims in a single blow, for the former committed the gravest sin in eating beef, and the latter became polluted by the fat of an unclean pig. Thus the sanctity of the cow helped fuel the Indian Mutiny against the British.

Even with the best intentions, it is sometimes impossible to uphold one's most sacred beliefs. The unfortunate Maharaja Sindhia of Gwalior was driving a train along a newly built railroad when a cow suddenly leaped onto the tracks and was struck dead. Years later the prince confided to a friend, "I think I shall never finish paying for that disaster, in penances and purifications, and in gifts to the Brahmins."

?

How did they start the *Guinness Book of World Records?*

It isn't Muhammad Ali or Mount Everest or the world's fattest woman who bears the distinction of inspiring the *Guinness Book of World Records*—it was a golden plover that started it all. In 1954 Sir Hugh Beaver was hunting in the green hills of Ireland. He shot down ducks and geese, but a flock of plovers escaped him. Gazing at the wintry sky, he wondered what had gone wrong. The birds must have been awfully fast—in fact, he told his friends that night that the plover was "the fastest game bird we've got." Sir Hugh may have salvaged his reputation of prowess in the hunt, but he did not get a consensus on just what the fastest game bird in the British Isles was. Nor did any encyclopedia provide the answer.

Sir Hugh, who was managing director of Arthur Guinness, Son and Company, the brewer of Guinness stout, went back to work in London but his thoughts were on a book, not beer—a book to which people might turn for a definitive record of all kinds of superlatives. The task seemed formidable, however, until one of his employees at the brewery came up with precisely the right people to tackle it.

All their lives the twins Norris and Ross McWhirter had been avid devotees of trivia. They had amassed hordes of facts and figures from newspapers and magazines and, having memorized a lot of it, they were particularly sensitive to vexing discrepancies. They organized a business in 1951 which would put the records straight and supply encyclopedias and newspapers with substantiated facts and figures on an enormous range of topics. When not clipping articles, the brothers wrote them: both were sports journalists and they happened to cover a runner who, being a junior executive at Guinness, put them in touch with Sir Hugh.

An interview took place (which Sir Hugh found most informative), and four months later the McWhirters had compiled and published the first volume of the *Guinness Book of World Records*, then only 198 pages long. Four months after that it was number one on England's nonfiction bestseller list.

Today, Norris McWhirter still edits the *Guinness Book of World Records* at Guinness Superlatives in Enfield outside of London. (His brother was shot and killed in 1975 by the Irish Republican Army.) Norris and his 14-member staff wade through some 20,000 letters a year from fans proffering new records, challenging old ones, inventing new fields of endeavor. Guinnessport, or the attempt to "get into Guinness," has become a popular craze, and McWhirter feels he has to set limits—barring alcoholic drinking contests, for example—to ensure the health and safety of his followers.

?

How did they discover champagne?

The next time you find an excuse to break open a bottle of bubbly—be it New Year's, a wedding, or a whim—you might raise your effervescent glass in solemn thanks to one blind Benedictine monk who made it all possible. Three centuries ago there lived in northern France a great blender of wines, Dom Pérignon, who served as cellar-master at the Benedictine Abbey of Hautvillers for 47 years. For some reason that we may never know, this monk decided one day to seal his bottles with cork instead of the usual cloth soaked in oil. The carbon dioxide that is produced during fermentation could pass through the cloth, but was imprisoned by the new stopper. The result: a sparkling wine.

Dom Pérignon is generally credited with having put the bubbles into champagne. This is stretching his achievement a bit, since growers for centuries must have noted that some wines referment in the spring, and the New Testament speaks of wine skins splitting with new wine. But Dom Pérignon put his observations to good use, strengthening the bottles and tying his thick corks down with string. The people of the Champagne region of France, at least, have not forgotten their inspired monk. An annual wine festival in Hautvillers celebrates his name and his achievements.

?

How did James Joyce support himself while writing *Ulysses*?

James Joyce began writing *Ulysses* in March 1914 at the age of 32. At that time he lived in Trieste with his wife, Nora Barnacle Joyce, and their two children in a flat furnished only with replicas of antique Danish chairs, perhaps emblematic of his love of Ibsen. His scant income was derived from two sources: he taught English in a public high school in the mornings, where he was noted for his tardiness as well as his eccentricity, then continued with English lessons on a private basis in the afternoon, in which he concocted elaborate stories to illustrate grammatical points. Among his students were Emma Cuzzi, sister of a prominent lawyer, and two of her friends, all age 14. Joyce would wrap up these lessons at the Cuzzi house by sliding down the banister, followed by his students, until Signora Cuzzi caught sight of them and ended the fun—and the lessons.

The outbreak of war interfered with Joyce's lessons and the public high school closed in the spring of 1915. Joyce's financial situation, which had always been precarious, became critical. He had no savings and still received no royalties for his literary works—on the contrary, eight years of litigation that preceded publication of *Dubliners* had set him back considerably. Since completion of his studies in 1902, Joyce had undertaken and quickly abandoned a career in medicine, written poems and "epiphanies" in Paris, returned to Dublin at the time of his mother's fatal illness, and left with 20-year-old Nora Barnacle for Zurich, where employment in the Berlitz School had been assured him by an agency in Ireland. When none was to be had, he headed for Italy—Trieste, Pola, then Rome, where he quickly found that corresponding for a bank was not his life's ambition. In 1907 he returned to Trieste, suffered rheumatic fever, and wrote "The Dead." Joyce traveled to Ireland in 1909 and 1913 to open the first regular cinema in Dublin and to battle for publication of *Dubliners*, returning to Italy with various of his sisters to relieve the poverty at his father's house.

While Joyce lived quietly and inconspicuously in Trieste, bored with his teaching and scarcely able to pay the rent, his work came (via William Butler Yeats) to the attention of a very influential literary figure: Ezra Pound. This energetic poet and outspoken herald of the avant-garde in the arts lived in London and was closely connected with two English reviews, *The Egoist* and *The Cerebralist*, as well as H. L. Mencken's *Smart Set* and Harriet Monroe's *Poetry* in America. Pound introduced Joyce's work to *The Egoist*'s editor, Dora Marsden, and her successor, Harriet Shaw Weaver, a shy feminist who became a champion of the brilliant Irishman and later a benefactress. Her and Pound's efforts to spread Joyce's fame in London and elsewhere were a boon to his spirits and financial situation.

Compelled to leave Triests in June 1915, the Joyce family made a hazardous journey through Austria and found asylum in Switzerland. In Zurich Joyce obtained some more language students and Nora's uncle gave them £25 on which to get started. Yeats and Pound then conspired to obtain a grant for Joyce from the Royal Literary Fund, enlisting the support of one of its members, Edmund Gosse. Pound's bold comparison of Joyce to Stendhal and Flaubert and Yeats's "conviction that [Joyce] is the most remarkable new talent in Ireland today" were not taken lightly, and a grant of £75 was soon on its way to the impoverished writer. He received, too, £2 a week for 25 weeks from the Society of Authors, and a year later a Civil List grant of £100 awarded by the Prime Minister as a result of patient and persistent efforts by Yeats and Pound.

Joyce's biographer, Richard Ellmann, comments that Joyce's years in Zurich were "marked by a great exfoliation of [his] creative powers. His talent grew in confidence and extravagance, dazzling and sometimes even disquieting to his friends." It was here he wrote most of *Ulysses* and established himself in the foreground of contemporary literature. Miss Weaver was an ardent believer in his genius and in February 1917 began to send him money anonymously, later followed by stocks that would afford a permanent income. This was augmented by the generous stipend of Edith Rockefeller McCormick, American patron of the arts, who was then living in Zurich and aided Joyce from the spring of 1918 through September 1919.

In mid-October, the war at an end, Joyce and his family returned to Trieste, where he resumed teaching at the commercial high school. The city was battered by war and the Joyce family's financial situation abruptly became critical once again. In 1920 Joyce resolved to travel north. He wrote Pound:

> I am in need of a long holiday (by this I don't mean abandonment of *Ulysses* but quiet in which to finish it) away from here. . . . I live in a flat with eleven other people and have had great difficulty in securing time and peace enough to write those two chapters. The second reason is: clothes. I have none and can't buy any. The other members of the family are still provided with decent clothes bought in Switzerland. I wear my son's boots (which are two sizes too large) and his castoff suit which is too narrow in the shoulders, other articles belong or belonged to my brother and to my brother-in-law. I shall not be able to buy anything here. . . . I can just live with what I have but no more. Since I came here I suppose I have not exchanged 100 words with anybody. I spend the greater part of my time sprawled

> across two beds surrounded by mountains of notes. . . . I could find lessons here (most people expected it of me) but I will not. I have a position in that school which the government has now raised to the rank of a university. My pay is about 3/- an hour for 6 hours a week. This I shall resign as it wastes my time and my nerves.

Pound caught up with Joyce in Italy and persuaded him that Paris was the place to launch his voluminous and unconventional novel. In July 1920 the Joyces arrived in Paris, having left Italy for the last time. Joyce already had many admirers in Paris, one of whom lent him a flat free of charge for four months; another supplied a bed and table, and a third gave him a warm coat. Joyce set about finishing *Ulysses* and finding a publisher. Those whom he contacted in England and America were reluctant. In April 1921 Sylvia Beach, owner of the renowned Shakespeare and Company,

James Joyce and Sylvia Beach at Shakespeare and Company in Paris; behind them, some reactions to the recently published masterpiece.

which still stands on the rue de l'Odéon in Paris, offered to publish the novel under the imprint of her shop. When Joyce readily agreed—not without dire warnings that no one would buy the book—Beach launched a heroic campaign on his behalf, aided by Adrienne Monnier, who owned a bookstore opposite her own. The two obtained support from everyone who came their way. Foremost among them was the writer and critic Valéry Larbaud, who announced, "I am raving mad over *Ulysses*," lent Joyce his apartment free of charge, and gave a public lecture about the novel two months before publication. Miss Beach planned to publish 1,000 copies and strove to obtain nearly as many advance subscriptions. Joyce would receive a generous 66 percent of the net proceeds (as compared with the average 10–15 percent most writers are rewarded with today).

Harriet Weaver took up the cause in England, sending Miss Beach a copious list of all bookstore owners and individuals who had expressed interest in the novel. She also sent Joyce £200 as an advance against royalties from the English edition. This, added to $150 a month supplied by American short-story writer Robert McAlmon (who blithely dipped into the vast wealth of his father-in-law, Sir John Ellerman), was sufficient to tide Joyce over, and allow him to go out drinking now and then at the Gipsy Bar with Wyndham Lewis or occasionally Valéry Larbaud.

Throughout the summer and fall of 1921 Joyce worked on completing "Ithaca" and "Penelope" while the galleys of earlier chapters began to arrive. He demanded five sets of the galleys and made enough changes—usually additions—to totally exasperate the most patient printer. Maurice Darantière in Dijon nevertheless persevered, and a weary Sylvia Beach faithfully acted on the stream of telegrams and letters with additional changes that poured in right up until publication.

The conductor of the Dijon express that arrived in Paris at 7 A.M. on February 2 was entrusted with the first two copies of *Ulysses*. One went to Sylvia Beach, the other to Joyce. It was his 40th birthday and, as he put it, "the expiration of my seven years' sentence." He appeared sad and tired rather than jubilant. In constant agitation he wrote to reviewers and friends to bring the book recognition. His life was still a shambles, with no permanent home. Nora was exasperated, and nearly half a year would pass before she even consented to read *Ulysses*.

?

How did the first European explorers cross North America and reach the Pacific?

In 1803 Thomas Jefferson secretly appropriated funds for exploration of the upper portions of the Missouri River and from there westward to the Pacific, a region not yet under the sovereignty of the United States. He asked his private secretary, Meriwether Lewis, and frontiersman William Clark to evaluate the territory for potential annexation, taking careful note of natural resources and trading opportunities. Lewis and Clark did span the continent in 1805 and, because of their significant role in America's westward expansion, many of us believe them to have been the first. Actually, Vasco Núñez de Balboa was the first European to travel from the Atlantic to the Pacific, but he did so by crossing the narrow isthmus of Central America in 1513. A yet lesser known figure who also preceded Lewis anClark was Alexander Mackenzie, who crossed Canada and reached the shores of the Pacific in 1793. This rugged Scotsman, one of a long line of fur traders, did it for wealth.

Sixteen-year-old Alexander Mackenzie came to Montreal in 1779. Ten years later he was in charge of Fort Chipewyan on Lake Athabasca in what is now Alberta, and he was intent on finding the Pacific. A first voyage took him north to the Great Slave Lake and along the Mackenzie River to the Arctic Ocean, but this cold, dangerous journey did not quell his spirit. In October 1792 he headed west from Fort Chipewyan to the farthest outpost of the fur settlers on the Peace River. The following May he set out from there with seven English and French settlers, and two Indians to hunt and interpret for them. All squeezed into a single canoe, just 25 feet long on the inside (i.e., excluding the curved bow and stern,) and 4 feet, 9 inches across at the broadest part. In addition, they loaded up some 3,000 pounds of arms and ammunition, baggage and provisions. One can only speculate how accurate an idea they had then of what was to come.

Mackenzie's ambition did not blind him to the wild beauty in which he found himself a lonely spectator. "This magnificent theatre of nature," he wrote like a true romantic, "has all the decorations which the trees and animals of the country can afford it: groves of poplars in every shape vary the scene; and their intervals are enlivened with vast herds of elks and buffaloes. . . ." Mackenzie not only noted the natural life around him, but areas alongside the Peace River where settlements might be made near the natural resources that would enable them to survive.

The canoe traveled virtually unharmed along the river to the foot of the Rockies, with some recaulking and occasional portages to get around rapids and falls. In swift currents where poling and paddling proved futile, the travelers unloaded provisions and towed the canoe. Eventually the river became a torrent of white water coursing the steep slopes from the mountains ahead. Again they were forced to haul out the canoe and continue on foot. To drag the boat they felled two rows of trees, but did not sever them completely from the stumps, so that, thus bent over, the trunks formed a railing on either side. Several men belayed ropes attached to the canoe around successively higher stumps, while others went ahead blazing a rough path through the wilderness. They continued like this for three days, exhausting their energies, fortifying themselves each night with rice and sugar and rum. After five weeks of journeying, in which they covered over 500 miles, they reached the Finlay River at the headwaters of many small streams. This was the great divide. Here they found small lakes and, along the shores, canoes, baskets, and fishnets belonging to Indians. Mackenzie helped himself to some of these, leaving a knife and beads as payment.

The stalwart group was now at the headwaters of the Fraser, a river that flows to the west. Mackenzie sent ahead some scouts to survey the waters, and they returned with ominous reports of boulders and frightful rapids. The explorers portaged the first part of the stream, but when they finally launched the boat again, they were swept rapidly downstream, then thrown against a gigantic boulder that smashed the stern. Men and provisions were flung into the seething white water. Gulping for air, flailing about for a solid handhold, they swam and scrambled ashore while precious cargoes of food and ammunition drifted out of sight. Exhausted, but at least alive, they rescued the boat and set about repairing it. Thereafter they continued on land, but huge fallen pines and dense undergrowth slowed their progress. At times they covered only a mile in five hours. When local Indians told Mackenzie of a path to the sea, now about 300 miles away (250 as the crow flies), he decided to hide the canoe and trek the remaining distance. Thus they proceeded, with packs weighing up to 90 pounds, Mackenzie further burdened by his large telescope. In mid-July they encountered Indians who said white men had been there many years ago. Mackenzie decided these must have been Captain Cook and his men coming upstream from the Pacific in 1776. Mackenzie followed the Bella Coola River until at last on July 19 he sighted an inlet of the sea: Queen Charlotte Sound, north of Vancouver Island. At Elcho Cove, on one of the channels leading into the sound, Mackenzie left an inscription written in vermilion and grease: ALEXANDER MACKENZIE, FROM CANADA, BY LAND, 22ND JULY 1793. Thereupon they turned around and went home, arriving safely at Fort Chipewyan a mere four months after they had left.

?

How did they decide that blue was for boys, pink for girls?

The assignment of a color to each sex serves the obvious purpose of distinguishing infants, who otherwise look pretty much alike. But the particular choice of blue dates back to ancient times, when evil spirits purportedly plagued young children but could be warded off with certain colors. Blue, emblematic of the heavens, had an immanent power to dispel satanic forces. (Many Arabs in the Mideast continue to paint their doorways blue.) Since it was of paramount importance to protect little boys, they were clothed in blue, while little girls were left to fend for themselves. Only much later did parents, somewhat guilty about the girls' lack of identity, assign them precious pink.

?

How did Dick Fosbury change the techniques of high jumping?

In Mexico's Estadio Olímpico, a crowd of 80,000 gasped as high jumper Dick Fosbury flew over the bar at seven feet, four and a half inches to win a gold medal in the 1968 Olympics and set an Olympic record. It wasn't only the height that astounded them (Russia's Valery Brumel still held the world record at seven feet, five and three fourths) but Fosbury's novel approach: head first and flat on his back.

Before jumping, Fosbury would stand at the start of the runway, sometimes for several minutes, meditating, worrying, visualizing himself clearing the bar. "I have to psych myself up," said the nervous jumper. "It's positive thinking, convincing myself that I'll make it." Then he'd bolt down the runway, just left of center, plant his right (outside) foot firmly parallel to the bar, and spring up, pivoting quickly so his back was to the bar, which he glimpsed behind, then beneath him from the corner of one eye. With his back parallel to the ground and crosswise to the bar, his legs dangled down on the starting side, till he jacknifed them up to clear the bar and land—appallingly enough—on his back or the nape of his neck. The usual pile of sawdust did not make for a welcoming landing base, so Fosbury finished his backward flight on three feet of foam.

Fosbury did not deliberately set about to revolutionize the world of high jumping, nor even to invent a sensational method for himself. The Fosbury Flop, as the jump is commonly called (christened by a sportswriter in Fosbury's hometown, Medford, Oregon), sort of evolved. "I didn't change my style," Fosbury told *Sports Illustrated*'s Roy Blount. "It changed inside me." Fosbury started jumping in fifth grade and was still using the scissors method in high school to clear five-four. (This is a sideways jump in which the athlete kicks up the leg nearer the bar, crosses in a sitting position, then brings the trailing leg up and over as the

Innovator of a novel technique, Fosbury "flops" over seven-one to win this high-jump event in the U.S. Olympic track-and-field trials in Los Angeles, June 30, 1968.

other leg comes down.) He tried the conventional straddle (crossing the bar on one's stomach, with one's body parallel to the bar) but it just didn't feel right, so he went back to the scissors. "As the bar increased," Fosbury recollects, "I started laying out more, and pretty soon I was flat on my back." He cleared five-ten at the time and the seed was planted for the Fosbury Flop.

Fosbury met with a lot of resistance from skeptical coaches along the way. It took a flying flop of about seven feet over a six-six bar to convince Fosbury's college coach Berny Wagner that the flop was more than a funny spectacle. "The physics of his jump are good," Wagner told *The New York Times*. "Dick exposes less of himself to the bar than any other high jumper."

Not only did Fosbury go on to glory with his backward, potentially hazardous leap, the technique became widely used among high jumpers. But the innovator himself said in 1968, "Sometimes I see movies, and I really wonder how I do it."

?

How did they get the idea that Shakespeare did not write Shakespeare's plays?

The shadowy actor of whom we know so little, with whom we credit the pinnacles of English literature, has been the object of veneration, gentle skepticism, and the most bitter abuse. The camps are divided, the theories myriad. Some worship the poet as a god; others denounce him as an illiterate, drunken rustic, usurer, and lousy actor.

A first, stealthy suggestion that Shakespeare was not the true author of his plays was made in the 18th century, but full-scale, international controversy did not erupt until the mid-19th century. It continues to this day, with well over 50 candidates proffered as the true Shakespeare. The debate has given rise to thousands of articles and books—in 1947 Professor Joseph S. Galland of Northwestern University compiled a bibliography on the subject running to more than 1,500 pages of manuscript, which no one could afford to publish. (The work has since been stored on microfilm.)

Most of the arguments against Shakespeare as the true author arise from a lack of information about his life, and the apparent incongruity of the genius of his plays with his unexceptional social position and limited formal education. It is not known whether Shakespeare went to university—reports of his education are scant—yet the plays are undoubtedly the work of one intimately familiar with the classics, the sciences, various languages, and the arts. Anti-Stratfordians (as the skeptics are called) focus especially on the professional knowledge of the law exhibited in the plays, the brilliant, supple use of legal language that could only have been provided by a lawyer. Other critics argue that the author of the plays must have traveled widely, to Spain, France, Italy, Denmark, and Scotland (there is no record that Shakespeare did), and have known well the aristocratic life within courts and palaces (Shakespeare was a commoner). Yet another theory is that if Shakespeare were the genius behind these laudable works, much would have been documented about his life, whereas, in fact, there is surprisingly little. Literary historians point to parallels in thought and style between Shakespeare's works and those of suggested candidates—Francis Bacon, Ben Jonson, and Christopher Marlowe, for example. And another camp has delved into a complex study of cryptograms—ciphers hidden within the text that purportedly reveal Bacon as the true author and, even more fantastic, that he was the son of Queen Elizabeth and the Earl of Leicester.

Even granting the skeptics their theories, one still wonders why the true author (assuming it was not Shakespeare) would go to such lengths to disguise his identity. The anti-Stratfordians are armed with a reply, which holds for most of the different theories of authorship: the noble lord who was furtively churning out these masterpieces could not have owned up to them without loss of stature. Although there were exceptions—Bacon included—generally it was not considered proper for an aristocrat to release poetry and plays to the public during his own lifetime. Furthermore, Shakespeare's historical plays inevitably took a political stance, and if it happened to anger the government, the author, claim the anti-Stratfordians, risked imprisonment or execution. *Richard II* did in fact enrage Queen Elizabeth, but no one was punished and this was a rare incident; her master of revels took care to censor all that came before her.

By the late 18th century, faith in Shakespeare had gone untainted for 200 years, and disbelief in his genius was nothing short of heretical. No wonder, then, that the author of the first outright statement discrediting Shakespeare subsequently burned his notes. This was the shy and retiring Reverend James Wilmot, rector of Burton-in-the-Heath near Stratford, who in 1785 concluded that Shakespeare lacked the necessary education to have written his plays. He went on to suggest Sir Francis Bacon as a more credible author. As a scientist, Bacon would have known about the circulation of the blood, alluded to in *Coriolanus,* whereas Shakespeare certainly would not. Wilmot noted, too, the names of three ministers at the Court of Navarre in *Love's Labour's Lost;* while Bacon's brother lived there for a time, we find no record that Shakespeare ever traveled there. Although Wilmot burned his notes, he confided his thesis to J. C. Cowell, whose records turned up in 1932.

In 1848 New York lawyer Joseph C. Hart, a colonel in the National Guard, kindled the debate with his strangely titled book *The Romance of Yachting.* Hart asserts that Shakespeare was "a vulgar and unlettered man," who secretly bought plays and put them on stage after lacing them with "obscenity, blackguardism and impurities." Hart named no other candidate, but the vituperative tone of his attack became characteristic of the strongest and largest group of iconoclasts: the Baconians.

In 1856 a frail New England lady named Delia Bacon (no relation to Sir Francis) published an article, "Shakespeare and His Plays: An Inquiry Concerning Them," the first of a torrent of articles culminating in her 543-page *The Philosophy of the Plays of Shakespeare Unfolded,* with a preface by Nathaniel Hawthorne. The fanatical Miss Bacon decried the tradition by which mankind was "condemned to refer the origin of these works to the vulgar, illiterate man who kept the theatre where they were first exhibited . . . condemned to look for the author of Hamlet himself—the subtle Hamlet of the university, the courtly Hamlet, 'the glass of fashion and the mould of form'—in that dirty, doggish group of players, who come into the scene summoned like a pack of hounds to his service." She looked beyond the crass "deer poacher" to "ONE, with learning broad enough, and deep enough and subtle enough, and comprehensive enough, one with nobility of aim and philosophic and poetic genius enough, to be able to claim his own, his own immortal progeny. . . ." This "ONE" became a group of Elizabethan scholars, of whom Bacon was only one. With him were Edmund Spenser, Sir Walter Raleigh, and several others, noble philosophers who carefully guarded their sacred gift of learning for the deserving and discerning, the ardent seekers of truth. Miss Bacon's arguments became increasingly convoluted and esoteric. Her admirers fell by the wayside, and the possessed spinstress sat alone in her unheated room in the home of a Stratford shoemaker, setting to paper her cryptic theories. While her book was being prepared for publication she haunted Shakespeare's grave, determined to prove her theory once and for all, for she firmly believed that buried with the poet were archives of the noble Elizabethan group. She went so far as to get the vicar's permission to open the grave, but at the crucial moment fears or doubts prevailed, and she turned away. Miss Bacon's hold on reality had perhaps been tenuous; with the publication of her book she became violently insane and ended her days in an asylum.

William Henry Smith was a somewhat more rational Baconian who in 1857 published a small book, *Bacon and Shakespeare*. His argument stems from our lack of knowledge of Shakespeare's life and from evidence of Baconian wit in the plays. The main platform of the Baconians, however, maintained throughout the 19th century and into the 20th, was the familiarity with law, both in terminology and in imagery, evident in the plays. Stratfordians and Elizabethan scholars counter that in the Tudor period there was a craze for the law, that people of all classes acted as their own lawyers and were well versed in legal procedures. Many other plays of the period show an equal, even superior, command of the field.

The Baconians went on to find parallelisms between Bacon's and Shakespeare's thoughts, and convoluted reasons why Bacon would publicly protest the theater while secretly writing for it. They hold up as "deadly evidence" the postcript of a letter from

Sir Tobie Matthew to Bacon in 1624: "The most prodigious wit that I ever knew, of my nation and this side the sea, is of your lordship's name, though he be known by another." Other miscellaneous items too numerous and esoteric to mention here add to the Baconian testimony, and some wild enthusiasts go so far as to credit Bacon not only with all of Shakespeare's works, but with those of Marlowe, Peele, and Kyd, as well as Spenser's *Faerie Queene* and Montaigne's *Essays*.

The cryptologists, who are by far the most bizarre members of the Baconian school, have made exhaustive searches for clues in the pagination and the designs of chapter headings and title pages. They find messages in the form of ciphers and anagrams concealed within the text by the true author who, some believe, wished to be known by posterity. All the leading systems of cryptology endorsed by the Baconians are outlined—and ultimately found lacking—in an authoritative work by William and Elizebeth Friedman, *The Shakespearean Ciphers Examined.* They review at some length the involved system of Ignatius Donnelly (1831–1901), an American lawyer and devout Baconian who believed that messages were to be found within the text of the plays. Working with the First Folio (which would have been available to Bacon), he scanned the text for a significant word. He then counted the number of words from the beginning of the first column to this word; one would expect the next key word to fall the same number of words after the first. But since this did not always work out, Donnelly tempered the system with "modifiers" or extraneous figures that could be added or subtracted at whim. He was inconsistent, too, in his treatment of hyphens and brackets. And when the problem really looked sticky, Donnelly supplied his own "subordinate root numbers" and computed until it worked out. In effect, what Donnelly did was find the message first—"Shak'st-spur never writ a word of them," for example—and then work backward. Proof of the arbitrariness of the system was made by the Reverend R. Nicholson, who, working with Donnelly's selected passage and following his system precisely, discerned *five times* the message "Master Will-I-am Shak'st-spurre writ the Play and was engaged at the Curtain."

Since World War I, Shakespeare skeptics have found scores of other champions. In many cases, their arguments parallel those that endorse Bacon. For example, J. T. Looney, an English schoolmaster, taught *The Merchant of Venice* year after year and concluded that its author must have had firsthand knowledge of Italy. Since Shakespeare failed, as far as is known, to meet this requirement, Looney searched for an Elizabethan poet of a similar style. He came up with Edward de Vere, 17th Earl of Oxford, who certainly knew well the kings and queens of England. The fact that no plays acknowledged to be by the Earl survive did not dampen Looney's spirits. He found incidents in the plays that might have been a part of the Earl's career, and thereupon published his theory in 1920. Professor Gilbert Slater subsequently expanded the authorship to include not only the Earl of Oxford, but the Earl of Derby, Francis Bacon, Sir Walter Raleigh, the Earl of Rutland, the Countess of Pembroke, and Christopher Marlowe. William Shakespeare, for once free of scornful abuse, was said to have been the middleman who negotiated with the theaters for production of the plays.

Popular in France is the claim made by a Frenchman in 1919 that William Stanley, the sixth Earl of Derby, was the true author. This nobleman was allegedly present at the Court of Navarre at Nérac in 1583, when certain events that appear in *Love's Labour's Lost* took place.

Finally, a more recent and in some ways the most outrageous claim is that made by Calvin Hoffman in 1955 for Christopher Marlowe. His argument resides in similarities between Marlowe's acknowledged works and Shakespeare's. That is all very well and is no surprise. But, according to the verdict of the Queen's coroner, Marlowe died in 1593 in a drunken brawl, long before many of the plays were written. Hoffman remains undaunted. We know the Privy Council was concerned about Marlowe's atheistic views, and that Marlowe had a wealthy patron, Sir Thomas Walsingham of Kent. From this and some prodigious digging, Hoffman speculates that Marlowe and Walsingham were lovers and that Walsingham, fearing for Marlowe's safety in light of government disapproval, staged the murder of a foreign sailor and bribed the coroner to name him as Marlowe, while the playwright was smuggled safely to Italy! There Marlowe continued to write; he sent his manuscripts to Walsingham, who arranged for their performance through none other than the humble actor William Shakespeare, who obligingly agreed to sign his name to them as well. Such are the lengths to which the discreditors of Shakespeare have gone.

The curious obsession with concocting theories to denigrate William Shakespeare has occupied lifetimes. Hundreds have pursued it passionately, fanatically, their reasons varied and not always clear. Ironically (for those of us who believe in William Shakespeare), this speaks for the genius of the poet, for nothing less could have inspired such fascination, ardor, and unrest.

?

How did Cardinal Richelieu, Prime Minister of France under Louis XIII, get his exercise?

By jumping over furniture.

?

How did they design the first car?

Four centuries ahead of his time, Leonardo da Vinci conceived workable plans for drawbridges and armored tanks, submarines, air conditioners, and machine guns. He imagined, too, the horseless carriage, a model of which stands in the museum in Vinci, about 20 miles west of Florence. The small, compact unit, constructed of wood, has three wheels under the body and a rudder bar attached to an additional small wheel in front for steering. Power is derived from a series of springs that operate two horizontal cogwheels, which drive the two large wheels of the car. In order to allow for the difference in speed of the two wheels as they go around a curve, Da Vinci placed a toothed crown wheel, which acts as a differential gear, between the two cogs. Hundreds

A car driven by springs, constructed from Leonardo's drawings.

of years later, when engineers carefully pieced together this vehicle from Leonardo's ingenious plans, they found it might work remarkably well for a spin through the hills of northern Italy.

?

How did they start the Baader-Meinhof gang—and how did the police catch them?

In 1972 Germany's "Bonnie and Clyde" were declared Public Enemies Number One. In the two years since its formation, the Red Army Faction, popularly called the Baader-Meinhof gang, had stunned, divided, and terrified the country. Over 100 police had been hospitalized and 1 killed, while judges' chambers were gutted and American installations fire-bombed. A new generation of fanatic youth had declared open war on bourgeois society. They robbed banks and stole cars (particularly BMWs); they raided a West German NATO munitions depot, amassing the most awesome private collection of ordnance in postwar Germany. As the years passed, their ambitions swelled and their victims increased. While much of the populace cried for law and order, liberals denounced that response; young people espoused the guerrillas' idealism, and the eminent writer Heinrich Böll shocked the world by comparing the search for Meinhof with the maniacal Nazi hunt for Jews.

Who were Baader and Meinhof, and how did they and their disciples come to endorse a dated Marxian ideology and flagrant violence to achieve it? Both Andreas Baader and Ulrike Meinhof were products of the system against which they fought, highly intelligent and well-educated middle-class citizens. Born in 1935 to two art historians, Ulrike Meinhof became a journalist for a Communist-backed paper, *Konkret,* edited by her husband, Klaus Rainer Röhl, and funded from Moscow. *Konkret*'s audience of students, agitators, and terrorists included just about everyone on the left except the proletariat they wished to mobilize. During these years, in the mid-'60s, Meinhof was also an affectionate mother of two, with a horror of guns. As Melvin J. Lasky pointed out in *The New York Times Magazine,* "Two souls, in Goethe's phrase, dwelt within Ulrike's breast. A struggle was taking place between Old Left earnestness and New Left liberation." In 1968 she made a radical and fateful move, slamming shut forever the doors to her past. She left her husband, devoted herself to the poor and oppressed with whom she had had no real contact, and took up the cry for revolution. But just before leaving, she published in *Konkret* an interview with Andreas Baader, who had been arrested for setting fire to two Frankfurt department stores, resulting in $700,000 in damages. Meinhof called his act "progressive," largely because it had defied the law.

Andreas Baader was a historian's son, art school dropout, and onetime member of "Red Rudi" Dutschke's Socialist Students' League. In 1970 Meinhof decided she needed him "to set up the urban guerrillas" and joined in a plot for his escape from jail. At this time, Baader was permitted to travel under guard to various Berlin libraries for his "sociological research" (he was allegedly writing a book about youth problems). On May 14 he was working in the reading room of the Free University when five figures, disguised in wigs, broke in, hurled tear gas, and opened fire. Several guards were wounded; all the guerrillas escaped. Andreas and Ulrike leaped from a first-floor window and dashed away in a silver Alfa Romeo to freedom and the birth of the Red Army Faction.

United in their ardor to immolate German society, their passion for Karl Marx and Herbert Marcuse, the two revolutionaries began forging passports, stealing cars, and robbing banks. With them was Gudrun Ensslin, Baader's "revolutionary bride," the zealous daughter of a Protestant pastor; and a hollow-eyed sociologist, Jan-Carl Raspe. Others who joined their forces were leftovers from various Marxist movements of the '60s, student and teacher activists, some of whom had been indicted in those years and could not make a living legally. By 1972 the group, numbering only about 25, had so jolted German society as to ignite massive retaliation. ". . . Police are carrying out the most extensive manhunt in West German history," reported *Time* magazine on February 7, 1972. "In Hamburg recently, more than 2,000 cops, using helicopters and dogs, sealed off all roads leading out of the city and conducted a twelve-hour search."

On May 19, 1972, two bombs wounding 17 people exploded in the Hamburg publishing house of Axel Springer, whose conservative papers had denounced the Baader-Meinhof gang. In the following two weeks the group was thought to be responsible for further bombings, including two at United States installations that left 4 dead and 41 wounded. Police maintained that the gang had connections with similar groups in Italy, Japan, and France. Meinhof had long overcome her distaste for guns, having received marksmanship training in a Palestinian guerrilla camp in 1970. Now, tens of thousands of police combed the countryside and posters of the leaders were seen everywhere.

At 5 A.M. on June 1, some 150 police surrounded a quiet house in north Frankfurt. An officer of the special squad of municipal, state, and federal police called through a bullhorn, "Come out, your means are limited but ours are unlimited." The house remained silent. The police threw tear gas grenades and the suspects opened fire. When the police started to ram the garage door with an armored car, a skinny young man stripped to his undershorts strode out calmly. This was Holger Meins, 30, a core member, but the prize was still to come. Andreas Baader suddenly burst from the house, shooting wildly. A policeman shot him in the hip and he was carried away on a stretcher, still shouting, "Pigs, pigs." The intellectual Jan-Carl Raspe was also seized. Six days later a Hamburg salesgirl tipped off the police that a customer in her fashionable boutique was browsing with a revolver sticking out of her purse. Police arrested Gudrun Ensslin, armed with two loaded pistols.

On June 15 a teacher in Hanover named Fritz Rodewald informed police of a series of mysterious guests who had come to stay with him, apparently "recommended by friends." Rodewald was among Meinhof's sympathizers who were providing safe houses for her supplies and followers. But he had come to the conclusion that her terrorism would ultimately hinder the New Left and so went out to a phone booth and notified the police. The ensuing raid turned up a cache of weapons, false identity papers, and a ten-pound bomb gift-wrapped and stashed in a red cosmetics case. The raiders also picked up Ulrike Meinhof.

The aftermath of the "Bonnie and Clyde" era was as disturbing and bizarre as its onset. Capture of the ringleaders by no means put an end to the Red Army Faction's activities. Germany was plagued by unpredictable and brutal acts of terrorism in the years that followed. Throughout West Germany and West Berlin, roughly 20 gang members sat in jails awaiting trial, but they did not sit idle. Meinhof wrote manifestos, while Baader devised intricate escape plots to be smuggled to the outside. Holger Meins died on a hunger strike in November 1974 after fasting for two months, emblematic of the group's fanaticism and a foreshadowing of further self-destruction. The next day gunmen shot and killed Gunter von Drenkmann, President of the West Berlin Supreme Court, and eight fire bombings occurred in Göttingen, all apparently in retaliation for Meins's death. On May 9, 1976, Ulrike Meinhof was found dead in her maximum security cell, hanged with a piece of prison towel. Prison authorities asserted suicide; Meinhof sympathizers naturally believed otherwise and said so with bombs in West Berlin, Paris, and Rome.

There remained at Stuttgart's Stammheim prison Baader, Raspe, and Ensslin, now in the midst of a two-year-long trial, one of the longest and most heated terrorist trials in history. In April 1977 the three were sentenced to life imprisonment for the killings of four American soldiers, but they were still to deliver a grisly message of their struggle.

In September industrialist Hanns-Martin Schleyer was abducted and his chauffeur gunned down. Among the kidnappers' demands was the release of 11 terrorists, including Baader, Raspe, and Ensslin. A month later, two Arabic-speaking men skyjacked a Lufthansa flight en route from Majorca to Frankfurt. The Boeing 737 was flown to Rome, Cyprus, Bahrain, Dubai, Aden, and finally Mogadishu in Somalia. Schleyer's captors claimed affiliation with the skyjackers, whose demands also included the release of the Baader-Meinhof ringleaders. Now the kidnappers, who were threatening to slay Schleyer, issued an ultimatum—"no more extensions." Chancellor Helmut Schmidt, firmly resisting the threats, ordered a dangerous and delicate operation. Twenty-eight commandos from West Germany's highly trained Grenzschutzgruppe Neun (Border Protection Group 9) skillfully reclaimed the Lufthansa flight, effecting the raid in a mere 11 minutes. Within a few hours of the rescue, amid the country's celebration and sighs of relief, Baader shot himself through the back of his head with a 7.65-millimeter pistol, Jan-Carl Raspe shot himself in the head with a 9-millimeter pistol, and Gudrun Ensslin hanged herself with an electrical wire. Told of the events, Schmidt retorted, "But that's impossible." All were in solitary confinement in a maximum wing at Stammheim. How could they have had such weapons? How could they have learned so quickly about the failure of the skyjackers? The director of the prison was fired and investigations began.

Two secret holes were discovered in the cells of Baader and Raspe. One could have held a pistol; the other contained batteries, wires, and sockets, the components of an elementary but astounding communications system. Authorities reasoned that Raspe had connected his telegraph system to the cell's thermostat and was able to tap out messages to the other terrorists, messages picked up from the outside on a tiny radio. Thus he would have followed the rescue mission in Mogadishu, learned with dread and anger of its success, and issued word to Ensslin and Baader to execute their suicide pact. Bewildered investigators speculated that prison guards must have known about the weapons (which also included knives and explosives), but had not dared to report them, for fear of reprisals against themselves or their families.

During the week of November 7, two very different groups mourned in the streets of Stuttgart. First came Schmidt and 700 official mourners at the funeral of Hanns-Martin Schleyer, whose body had been retrieved from the trunk of a car in Mulhouse. Two days later, 1,000 anguished radicals stood at the graveside of the terrorists. Both inside and outside Germany, demonstrators kept alive the fires of terrrorism with political slogans, bombings, and banners: GUDRUN, ANDREAS, AND JAN—TORTURED AND MURDERED AT STAMMHEIM.

?

How did they lay the first transatlantic cable?

For thousands of years, communication moved only as fast as a horse could run or a ship could muster subject to the vagaries of the wind. But in the 19th century, an invention popularly attributed to Samuel F. B. Morse suddenly transcended the miles, within countries and beyond. Miraculous wires began to creep over Europe and the more settled regions of North America. By mid-century the telegraph system was proliferating rapidly, defying distance and shrinking the vast earth, which would diminish further with the telephone, automobile, and airplane. In 1850 a telegraph cable was laid across the English Channel and was promptly followed by a link between Dover, England, and Ostend, Belgium; four between England and Holland; and a spanning of the Black Sea in 1855 to enhance British communications in the Crimean War.

Morse had predicted that one day the Atlantic would be tackled, but this momentous feat awaited a champion—one who would doggedly raise funds, convert the skeptics, and maintain faith throughout years of defeat. This was Cyrus W. Field, son of a New England Congregational minister. Field left home at the age of 15 with $8 in his pocket and went to New York to seek his fortune, which within a relatively short time he found. He might have ended his days quietly touring Europe with his family had he not met F. N. Gisborne, an English engineer of the Newfoundland Electric Telegraph Company, then immersed in plans to lay a line across Newfoundland. Although the scheme flopped and the company went bankrupt in 1853, a seed was firmly planted in Field's mind, a dream to join two continents, to communicate through a wire sunk two miles beneath the black and silent sea.

Over the next five years Field bailed out the Newfoundland Company, obtaining the support of Chandler White, Moses Taylor, Peter Cooper, and other optimistic capitalists. In England he met English engineer John Brett, responsible for the cable across the Channel, and a brilliant young engineer, Charles Tilston Bright, who at 24 became chief engineer of Field's operation. Edward Whitehouse was enlisted as the company electrician. Field then turned to the British government, which, encouraged by Morse's recent announcement that a signal could indeed be transmitted over 2,000 miles, agreed to supply £14,000 per year. More capital was provided by Lady Byron and William Makepeace Thackeray, and in 1856 the Atlantic Telegraph Company was born with £350,000 to its name. The American government, however, was hardly enthusiastic, objecting to the cost, the close contact with England, and the interference of the state in private business. Finally, out of the tangled controversy a bill emerged providing income and ships, and it squeaked through the House on March 3, 1856, by a single vote.

Production of the cable was already under way in England, as Field had assured his backers the job would be done by 1857. In a mere six months 335,000 miles of iron and copper wire was drawn and spun, then covered with 300,000 miles of tarred hemp; this was shielded by strands of twisted iron wire. The result: 2,500 miles of cable, 500 miles longer than the distance from Ireland to Newfoundland. (The extra length provided slack in laying the line and a supply in case of damage.) The total weighed 2,500 tons, a formidable amount for any ship to bear. It was therefore divided between two warships, the United States' *Niagara* and

Britain's *Agamemnon*. Space was carved out in the hulls to accommodate the cable, but the *Agamemnon* still had to store large quantities on deck. On August 5, 1857, the two ships arrived at Valentia Bay, County Kerry, Ireland, and one end of the cable was landed. The following day they turned their sights westward, the *Niagara* to lay her share and then, in midocean, splice it onto the cable aboard the *Agamemnon*. The *Niagara* edged away at two miles an hour, with the cable paid out slightly faster to compensate for the uneven ocean floor. Communication with the shore was kept up at all times except on the fifth day, when the line mysteriously went dead for a few hours. The following day, the men had to tighten the brake on the paying-out mechanism, and in the process the cable was strained and snapped. The downcast crew returned to England with an unused 2,200 miles of cable.

On board this expedition and subsequent ones was a distinguished gentleman who played an essential role in the planning and laying of the cable: Professor William Thomson, Lord Kelvin, one of the greatest scientists of his century. He immediately began studying the shortcomings of the cable, which were many, partly due to the hastiness of the venture. Among other improvements, he proposed a sensitive detector at the receiving end that would allow a higher rate of signaling. The markedly less brilliant Whitehouse disagreed and argued for sheer volume and power in the transmission of the cable.

In the spring of 1858 the two ships set out again, loaded with new cable. This time they would meet in the mid-Atlantic, splice the cable, and head for their respective shores. But two days out at sea, the *Agamemnon* was hit by a ferocious gale. "The massive beams under her upper deck coil," wrote Nicholas Woods of the London *Times*, who was on board, "cracked and snapped with a noise resembling that of small artillery, almost drowning the hideous roar of the wind as it moaned and howled through the rigging. . . . Four gigantic waves were seen approaching the ship, coming slowly on through the mist nearer and nearer, rolling on like hills of green water, with a crown of foam that seemed to double their height. The *Agamemnon* rose heavily to the first, and then went lower quickly into the deep trough of the sea, falling over as she did so, so as almost to capsize completely.

There was a fearful crashing as she lay over this way, for everything broke adrift. . . . [A] confused mass of sailors, boys, marines, with deck-buckets, ropes, ladders and everything that could get loose, were being hurled in a mass across the ship. . . . *The lurch of the ship was calculated at forty-five degrees each way for five times in rapid succession*. . . . The coil in the main hold . . . had begun to get adrift, and the top kept working and shifting as the ship lurched, until some forty or fifty miles were in a hopeless state of tangle, resembling nothing so much as a cargo of live eels. . . ."

Miraculously, the ship and crew survived and the rendezvous with the *Niagara* went according to plan. But three times the ships tried to part and three times the cable failed, breaking when they were 3, 80, and then 200 miles apart. Disheartened and low on provisions, the men returned to Ireland. Holding fast to the feasibility of his dream, Field had the ships out again in midsea that same summer, splicing the cable for the fourth time. The *Niagara* enjoyed an uneventful few weeks calmly churning toward Newfoundland. The battered *Agamemnon* sighted a whale agonizingly near the cable, suffered an inexplicable failure of the conducting core (and then resumption of power), and another relentless gale. With coal supplies nearly depleted and a crew at once exhausted and excited, the hardy ship approached Valentia on August 5. Communication still crackled through the wires embedded in the mud and rocks of the sea, and celebrations began. On August 16 Queen Victoria sent a message by cable to President Buchanan. Fireworks exploded, and Charles Bright was knighted.

The Queen's greeting had, however, taken 16½ hours to transmit. As Arthur C. Clarke explains in *Voice Across the Sea,* Thomson and Whitehouse were disagreeing over the voltage of batteries used in sending signals and the type of equipment employed in receiving them. Whitehouse's insensitive automatic recorder failed to detect messages at one end. Thomson, on the other hand, had solved the problem with his mirror galvanometer, conceived one day as he noticed the play of sunlight from his monocle; this gave him the idea of a weightless pointer. On the galvanometer, the extremely tiny deflection of a coil carrying an electric current was magnified by a point of light reflected from a small mirror attached to the coil. The debate raged but meanwhile, as a result of poor insulation, the costly cable died. The public was outraged, calling the whole thing a hoax and a fraud. Poor Cyrus Field was out of pocket the entire £350,000 and it would take seven taxing years of hard talking before he got the backing to do it again.

Although the delay may have been frustrating, it did allow for long-overdue research into the cable's design. The result was a vastly improved, more substantial cable, with a conducting core three times as large as that of 1858, and a breaking strain of eight tons as compared with three tons for the earlier one. With the improvements came still greater weight: 2,600 miles of cable completed in 1865 weighed a staggering 7,000 tons. But now Field found a ship—the only one in the world—that could hold the entire length. This was Isambard Kingdom Brunel's magnificent *Great Eastern,* 700 feet long, with six masts, a displacement of 32,000 tons, and surprisingly supple maneuverability. A smaller ship landed the cable via a bridge of boats, at Foilhommerum Bay, five miles from Valentia Bay, while festive crowds looked on from the surrounding hills. The cable was spliced on board the *Great Eastern,* waiting out at sea with her 500-man crew, Field and Thomson, and three giant tanks for the cable. On July 23, 1865, she headed west. At a distance of 84 miles the crew detected an electrical fault and began hauling in the cable with wire tackle. Ten miles back, they found a most alarming problem: a two-inch piece of wire had been driven right through the cable. It certainly looked like sabotage, and the problem recurred five days later. Thereafter, a guard was stationed in the tank room to oversee the workers.

With three fourths of her journey behind her and 1,300 miles of cable safely laid, the *Great Eastern* suddenly disappeared. The sharp, clear signals it had issued steadily to England stopped short. Days turned into weeks. All imagined the worst. Confusion and disillusion prevailed, but in the midst of it all the Atlantic Telegraph Company brazenly began seeking money for a second cable, while uttering assurances that the first was proceeding as planned. As it turned out, the *Great Eastern* was still afloat, but the cable was not—severed and lost 2,000 fathoms deep. A fault had again been sighted and the cable was, as before, cut and hauled on board. But the machinery failed, and the wind shoved the ship around so that the cable scraped against her side. The strain of heaving it up was so great that the cable suddenly snapped and crashed back down into the water, sinking calmly from sight.

The determination of those men on board to complete their task is clearly seen in their long and tantalizing search to salvage the cable. For 11 days they scraped the bottom with a five-pronged grapnel attached to five miles of wire rope, frequently catching but never landing their elusive game. Each time shackles split, for the strain was too great, and the cable slipped back. In a heavy fog the ship drifted 46 miles from the cable and had to reposition in bad weather and unfavorable winds. In the final attempt, the hawsers snapped and two miles of iron coils and wire went down with the cable to the slimy bed of the mid-Atlantic.

Once again Field turned to fund raising. This time the public displayed some confidence that the thing could be done, and Field managed to float a new company, the Anglo-America Telegraph Company, and raise £600,000. He ordered 1,990 miles of new cable and replaced the brittle iron covering with a more ductile type of galvanized iron. At last the *Great Eastern* was supplied with machinery that could haul in the cable as well as pay it out, and electrical testing was implemented that could detect faults as the cable went out.

In a year the *Great Eastern* was ready to sail, slipping gracefully out of Valentia Bay on July 13, 1866. Fourteen days later the crew sighted land and anchored, aptly enough, in the serene waters of Heart's Content, an inlet of Trinity Bay, Newfoundland. The long-awaited cable was in place and in superb working order. The first day of operation brought in $5,000.

One would think Cyrus Field had seen enough cable to last a lifetime, but even now with success in hand, he would not go home. He had to go back and get the broken cable, which lay dormant, like some age-old serpent on the bottom of the sea. In August of the same year, the *Great Eastern* was back at the site of the previous year's defeat, which had been marked by a buoy. A grapnel attached to a two-inch-thick wire rope combed the ocean floor, back and forth, groping blindly through rocks and mud. It took a week to hook the cable, and the strain of lifting caused it to snap and fall back. This happened repeatedly. At the end of three weeks, the men again hooked the cable and managed to buoy it at one point while grappling it at a second, thereby dividing the strain. At last they had their enormous catch, but there remained the awful possibility that it had died over the previous year. For the electrician in England who had stood by the cable for a year, the first message must have seemed a miraculous resurrection of the dead. The cable, alive and healthy, was spliced onto additional cable on board. Messages darted to England and from there via the completed cable back to America, so that all could follow the *Great Eastern*'s finale. By the end of August two cables spanned the Atlantic. Thomson received his knighthood and Field a gold medal, overdue thanks from Congress, and cheers from the public, welcoming a new age.

?

How did they know the earth is round?

Aristotle provided the first conclusive argument for a spherical earth when he noted during a lunar eclipse that the shadow the earth cast on the moon was circular.

?

But how did they know the earth is flat?

Lactantius of Alexandria was an early apostate from the scientifically based cosmology of the ancients. In "On the False Wisdom of the Philosophers," a chapter in his copious *Divine Institutions* (A.D. 302–323), he ridicules the idea of a spherical earth in the most simplistic and nearsighted terms. The devout author scoffs at the image of people with their feet above their heads, rain falling upward, mountains hanging on without support; and he flatly denounces the possibility of heavens *lower* than the earth.

The Fathers of the Church championed increasingly narrow and literal interpretations of the Scriptures, deriving the shape of the earth and motion of the planets from Genesis and fragmentary passages throughout the sacred text. (The Huns who overran Europe and toppled the Roman Empire lent a helping hand to this effort to extinguish classical learning.) Severianus, Bishop of Gabala, a leader of the Syrian Church in 360, held the heavens to be a tabernacle—a theory popular among many patristic writers—on the basis of Isaiah 40:22: "It is he . . . that stretcheth out the heavens as a curtain and spreadeth them out as a tent to dwell in. . . ." Severianus taught that the sun does not pass under the earth at night but travels through northern parts, as if concealed behind a wall. "The sun also ariseth, and the sun goeth down, and hasteth to his place where he arose" (Ecclesiastes 1:5). In winter, he argued, the days are shorter because the sun takes longer to travel its nightly course.

Although the Fathers of the Church never established an organized and detailed cosmology, a well-traveled merchant named Kosmas Indicopleustes wrote 12 books on the subject between 535 and 547. As part of his elaborate theory of the universe as tabernacle, in which the earth appears as a rectangular table of shewbread (loaves that were offered to the Lord every Sabbath), he points out that this plane is higher in the north and west, for ships traveling in this direction are sluggish compared with those sailing south and east. Similarly, the Tigris and Euphrates flow along more readily than the Nile. Kosmas maintains that at night the sun goes behind a mountain—near the top in summer, resulting in a short night; along the base in winter, bringing long hours of darkness.

Hrabanus Maurus, Abbot of Fulda and later Archbishop of Mainz, decided in the 9th century that the earth was round, like a wheel, but he had to reconcile this with the Scriptures that speak about the four corners of the world. The solution for him was to put the square inside a circular horizon, and give heaven two doors through which the sun could pass. By this time, however, the Greek and Roman writings had again gained credence and were moving westward, overturning the tables of the tabernacle, making way for the gods of science.

?

How did they break the Japanese secret code in World War II?

By the mid-1930s the U.S. Navy had begun to solve some Japanese diplomatic messages with relative ease. But just as the Navy succeeded in breaking what the United States called the Red code, Japan's Foreign Office installed new, more sophisticated coding machines in its major embassies, and the ensuing transmissions had the Navy baffled. The new Japanese system, called Purple by the Americans, was reserved for top-secret diplomatic communications, and it was one of the most complicated devised before the age of computers. In order to break it, American officials initiated the greatest effort to date in the field of cryptanalysis, the science of breaking codes without prior knowledge of the keys.

The responsibility for solving the Purple system was shared by the Navy's OP-20-G, its radio intelligence section in Washington, D.C., and the Army's Signal Intelligence Service, or SIS, also in Washington, an organization headed by the civilian cryptologist William F. Friedman. A Russian-born geneticist who had become interested in cryptology at a research laboratory in Geneva, Illinois, Friedman had come to be recognized as the premier codebreaker in the United States. The Army began intercepting Purple messages in 1937, but it was not until February 1939, when Chief Signal Officer General Joseph O. Mauborgne, Friedman's boss, urged Friedman to drop all of his administrative obligations and devote his full attention to Purple, that a major attack was mounted. Friedman understood that his work, along with that of the team he directed, would be conducted under total secrecy; even his wife, Elizebeth, herself a cryptologist, would not know the nature of the project. As the painstaking task progressed, Friedman would find his personal involvement increasing to the point where his nights as well as his days were filled with thoughts of nothing but the Purple machine.

Friedman's first step in solving was to collect messages sent on the same day, since presumably all of these messages would be controlled by the same keys. Next, his team looked for patterns, such as repeated letters or the absence of certain letters, in an attempt to determine the rules governing the system. Knowledge obtained from breaking of the earlier, Red code enabled the team to identify paragraphs, which were usually numbered, and standard phrases such as "I have the honor to inform Your Excellency . . ." Newspaper dispatches often provided clues to the subject matter of messages, and sometimes translations of entire cryptograms were received when the Japanese embassy submitted their dispatches to the U.S. State Department. On rarer occasions, whenever a Japanese coding operator had made a mistake, the same message would be sent twice. The codebreakers would then have two nearly identical, or isomorphic, messages in front of them, and by locating the discrepancies they could zero in on the keys to the code.

Progress was sporadic. But gradually, with the help of mathematicians using group theory and congruences, Friedman's team, assisted by the Navy, put together a sketch of the Purple system, a kind of cipher that could be illustrated by two alphabet tables, one 20 consonants wide by 20 deep, the other 6 vowels wide by 6 deep. (The Japanese transliterated their language into Roman letters for encrypting.) From their pencil-and-paper diagrams, the team began to build a model of the Purple machine.

Their inelegant contraption consisted of two typewriters connected by wires to a drawer-sized coding box between them. When a message was typed in plain language on one typewriter, it would automatically be sent through the box and emerge in coded form on the other typewriter. The nerve center of the machine was four telephone selector switches within the box. These switches took the electrical impulse caused by a key's being depressed on the input typewriter and shifted it to another letter on the output typewriter. They produced hundreds of thousands of combinations before arriving back at their starting positions. Any letter of the alphabet could be represented by a different letter each time it was enciphered. *A*, for example, could be represented by *J* the first time it appeared, *T* the second, and *R* the third, all within the same message.

What the American team had constructed was a replica of what the Japanese called their 97-shiki O-bun In-ji-ki, or Alphabetical Typewriter '97. The '97 stood for the Japanese calendar year 2597, corresponding to 1937. The Japanese, in turn, had adapted their version from a German prototype called the Enigma machine. To encode a message, a Japanese operator would look up the key for the day in a thick Yu Go key book, plug in the 26 wires leading from each typewriter according to the key, and begin to send. If he so desired, the operator could superencipher an important communication by first encoding it in a code book and then enciphering the code text. The American model resembled its Japanese counterpart almost exactly, down to the loud humming noise it made and the showers of sparks it sometimes emitted as it ran.

Using their machine, Friedman's team read the first complete Purple message in August 1940. Shortly after the initial breakthrough, Lieutenant Francis A. Raven, a member of the naval team of codebreakers, discovered the key to the keys. He found that the Japanese had divided each month into thirds and that the keys for the last nine days in each ten-day period were jugglings of the key for the first day. Solutions, then, were needed for only the 1st, 11th, and 21st days of the month. But although some master keys were identified in minutes, others required weeks or months to locate, and still others proved so elusive they were never recovered.

Following their success, the SIS built several copies of the Purple machine and distributed them to the Army, the Navy, and certain British codebreaking groups. The British, who had been stymied in their attempts to break Purple, partially returned the favor by passing along a little information about German codes and ciphers.

In October 1940, after 18 months of intensive work on the Purple project, Friedman was made an Army Reserve lieutenant colonel, but less than 2 months later he collapsed from nervous exhaustion and was taken to the psychiatric ward of Walter Reed Hospital. Soon after he left the hospital in March 1941, he was honorably discharged from the Army for medical reasons. He continued to work in the SIS, as a civilian, for the duration of the Second World War.

On December 7, 1941, at 5 A.M., the SIS broke the last section of a 14-part ultimatum sent from Tokyo to the Japanese embassy in Washington. This last part contained a veiled declaration of war, saying that Japan was breaking off relations with the United States, and that the entire 14-part message was to be delivered to U.S. Secretary of State Cordell Hull at 1 P.M. that afternoon. The men who intercepted the message correctly assumed that the time of delivery would correspond to a Japanese military attack on the Allied Forces somewhere in the Pacific. But where, exactly, would the attack occur?

One P.M. Eastern Standard Time corresponds to dawn in Hawaii, but Washington's efforts to notify Pearl Harbor in the quickest possible manner were frustrated by a series of blunders. President Roosevelt thought it more likely that the Japanese would attack the British in Malaysia than the United States Pacific fleet. Moreover, American military officials feared that a direct radiotelephone call to Hawaii could alert the Japanese—who would be listening in—to the fact that the Purple code had been broken. Therefore, the news of an impending attack was sent by telegraph. Army Chief of Staff General George C. Marshall and other officials believed it would go through military channels; however, due to a mistake by communication personnel, the message was passed on to San Francisco by commercial telegraph, relayed from there to Hawaii, and finally delivered by motorcycle messenger to an American commander hours after the Japanese strike force had arrived.

Even though the breaking of the Purple code failed to avert the Pearl Harbor disaster, later interceptions of Japanese messages aided the Allies enormously. They were invaluable, through the reading of messages of Japanese diplomats in Berlin, in revealing many of Hitler's plans. As General Marshall once wrote, "The conduct of General Eisenhower's campaign and of all operations in the Pacific are closely related in conception and timing to the information we secretly obtain through these intercepted codes. They contribute greatly . . . to the saving in American lives, both in the conduct of current operations and in looking towards the early termination of the war."

?

How did they pick the Four Hundred?

If one had stood at the corner of New York's 34th Street and 5th Avenue at eight o'clock on a given night in the 1870s, one might have seen a string of stately carriages arrive at Number 350 on 34th. There the crème de la crème of New York Society would descend in a glitter of diamonds and swish of silk, poised, self-righteous, and urbane. These were the elite, critically judged for birth, background, and breeding—wealth alone was certainly insufficient—who had won the honor of dining with Caroline Astor, the commanding leader and arbiter of high society. The guests were greeted by Mrs. Astor herself, regally stationed beneath an immense portrait of herself in a velvet gown trimmed with lace and diamonds. In the oak-paneled dining hall, an army of servants circulated for three or four hours, serving a ten-course feast on gold plates, reputedly worth $400 apiece. The tables moaned beneath a profusion of roses and orchids spilling from their golden bowls. Fine French chefs were imported to concoct the most lavish gastronomical delights—soup of the rare *tortue*, beef larded with truffles, a confection of sweetbreads, pâté de foie gras, the essential canvasback duck, sorbets, French cheese, bonbons, and gateaux. This was the Gilded Age, the years of Patriarch Balls and dinner after the opera at Delmonico's. Junior dancing classes afforded an opportunity for the proper young to meet and so perpetuate the purity of the aristocracy. Nothing in life could be more significant than high society, and those who were a part spent millions to stay there.

Caroline Astor (née Schermerhorn) was born into a wealthy

Mrs. William Astor stood beneath this regal portrait of herself by Charles Émile Carolus-Duran as she received guests at dinners and balls in the 1890s. The shoulders and cuffs of her velvet gown are studded with diamonds, but Mrs. Astor generally adorned herself as well with a staggering weight of precious stones.

mercantile family, which claimed descent from noble Dutch patroons. At 22 she married William Backhouse Astor, Jr., second grandson of the original John Jacob Astor, who arrived in this country a ragged German immigrant in 1784 and by 1841 was the richest man in America. Although William was bright and well educated, primogeniture held sway in the Astor household, and he always took the back seat to his brother, John Jacob III. Caroline's domineering character was a further cause of William's dismay—she even made him drop his ignoble middle name—and, after fathering four daughters and a son, he took to the seas as a peripatetic millionaire playboy on his extravagant yacht, leaving Caroline free to carry on at home.

Carry on Mrs. Astor did, with the loftiest ambitions. Whether to ensure the most proper marriages for her children or for the sheer love of power, prestige, and the exercise of them, she deliberately set about becoming Queen of Society. No one ever knew what passed through Caroline Astor's mind. She maintained a forceful yet inscrutable presence amid her exclusive entourage of bluebloods. But Caroline had some assistance in her ascendancy; her majordomo and guide on everything from *pavé de boeuf* to guest lists was a paunchy, pompous fellow with a scraggly Van Dyke beard. Ward McAllister, a southern lawyer, found far more interest in champagne and balls than in the law. He aspired to a place in New York Society and wanted nothing less than to commandeer it. Beginning in the 1850s he gave the most magnificent gourmet dinners, which rapidly drew the attention of wealthy gentlemen. He organized balls and cotillions; he studied family trees, and pronounced New York Society a bit too amorphous. By the late 1860s McAllister decided it was he who would trim and shape it. He approached the most snobbish and wealthiest men in New York and together they decided "to band together the respectable elements of the city [so that] the good and wise men could easily control Society." McAllister composed a list of 25 so-called Patriarchs, gentlemen with a requisite four quality generations behind them. Their purpose was to give balls; each man took on the grave responsibility of inviting four suitable ladies and five suitable gentlemen. McAllister's successes were many, but the pinnacle of bliss came when Caroline Astor—*the* Mrs. Astor—sought his guidance. Together these two formulated

THE ONLY FOUR HUNDRED

WARD M'ALLISTER GIVES OUT THE OFFICIAL LIST.

HERE ARE THE NAMES, DON'T YOU KNOW, ON THE AUTHORITY OF THEIR GREAT LEADER, YOU UNDERSTAND, AND THEREFORE GENUINE, YOU SEE.

"The so-called Four Hundred has not been cut down or dwindled to 150 names," said Ward McAllister yesterday. "The nonsense, don't you know, printed to that effect in the *World* and some other papers, has made a very bad impression that will reflect badly against them, you understand. That list of names, you understand, printed on Sunday, did not come from me, don't you see. It is unauthorized, don't you see. But it is accurate as far as it goes, you understand.

"It is incomplete and does injustice, you understand, to many eligible millionaires. Think of leaving out such names, don't you know, as Chauncey M. Depew, Gen. Alexander S. Webb, Mr. and Mrs. Edward Cooper, Mr. and Mrs. Luther Kountze, Mr. and Mrs. Robert Goelet, Mrs. and Miss Wilson, Miss Greene, and many others! Don't you understand, it is absurd, senseless.

A statement from Ward McAllister, reported tongue in cheek by the press.

fashions and carved in gold just who and what was accepted, for over 20 years. Despite Mrs. Astor's stout physique, bulbous nose, and homely looks, McAllister called her Mystic Rose in the troubadour tradition of courtly love. And McAllister—dubbed Make-a-Lister—was creator of the term "Four Hundred" and critical judge of who got on that list.

It began one day in the early 1890s, when McAllister casually remarked to the press that there were only about 400 people in fashionable New York Society. "If you go outside that number," affirmed Ward, "you strike people who are either not at ease in the ballroom or else make other people ill at ease." And when in 1892 Mrs. Astor gave a final ball in her mansion on 34th Street, McAllister asserted that the ballroom could accommodate no more than 400. (Given the enormous hoop skirts then in fashion, he was undoubtedly right.) He and Mrs. Astor then undertook the delicate and dangerous task of trimming the list of possible guests. Their touch could make or ruin. McAllister did not let his list be known all at once; rather, he taunted the public by revealing pertinent information a little at a time and terrified society by classifying even the privileged Four Hundred. The Original Inner Circle, of which he was one, comprised 150; there followed 19 in the Contingent Inner Circle Margin, 26 Star Members Inner Circle Fringe, 49 called Plain Inner Circle Fringe, and 156 designated Fringe to Plain Inner Circle Fringe.

How were the selections made? First, Mrs. Astor had to know each member personally, and this excluded many who were quite eligible. McAllister perused his family trees, finding descendants of the old colonists as well as families who had achieved social prominence a bit more recently. The entire charade was particularly controversial since Mrs. Astor's own position derived from comparatively recent wealth. The list included such Old Guard names as Livingston, Cushing, Jay, Cruger, Chanler, and Fish, and some later additions, Vanderbilt and Mills. There were some novel and very puzzling inclusions on the guest list, too: the Mayor of New York City, an editor of the *New York Sun,* two architects, and a sculptor.

On the evening of Mrs. Astor's gala event in 1892, the mansion was ablaze and the hostess as usual "borne down by a terrible weight of precious stones." Mrs. Astor had a seemingly limitless collection of gems and her philosophy was to wear as many as possible at one time. With diamonds dripping down her back and glittering from pompadour to midriff, she sat on a large divan on a dais and surveyed the scene, calling a lucky few to chat with her. This was her apotheosis, and Ward McAllister was made a celebrity as well. Melville Publishing Company decided to publish his list, *McAllister's Four Hundred: Officially Supervised,* with a curious introduction by the social leader himself: "Our catalogue has been prepared with much care, the names having been well sifted and weighed, and only those admitted who are now *prominently* to the front, who have the means to maintain their position, either by gold, brains, or beauty, gold being always the most potent 'open sesame,' beauty the next in importance, while brains and ancestors count for very little."

Whether he was being glib or had been converted to the ever more powerful influence of the dollar is uncertain. But he had made his mark and left the general public believing that the size of the Astor ballroom determined the composition of society.

?

How did they discover the atom is mostly empty space?

For a long time the exceedingly small size of an atom—one hundred-millionth of an inch is typical—made it very difficult to make even an informed guess about its structure. Sir Isaac Newton imagined atoms as "solid, massy, hard, impenetrable, moveable particles." Others decided they were red, or gray, compact, and densely populated with positive and negative electrical charges.

In 1911 the eminent physicist Lord Rutherford performed an experiment whose results astonished him and broke ground for the nuclear age. He had his assistant, Hans Geiger, shoot a beam of rapidly moving alpha particles at a piece of gold foil. The speed and impact of these alpha particles should, according to Rutherford, have carried them right through—and for the most part that was what occurred. But on closer examination, Geiger found a small number of alpha particles on the side of the foil on which they had entered. "It was almost as incredible," exclaimed Rutherford, "as if you fired a 15-inch shell at a piece of tissue paper and it came back and hit you."

It wasn't long before Rutherford came up with an explanation. Some of the alpha particles came back because they ricocheted off a hard substance. But since so few of the particles came back and since most could continue through, Rutherford figured the hard substances must be very tiny and very far apart. Thus, the majority of each atom must be empty space, and only a much smaller part occupied by compact matter. This extremely small center Rutherford called the nucleus.

Physicists set to work to explore the atom further, discovering the electron in 1914, the proton in 1920, the neutron in 1932. These finds confirmed Lord Rutherford's conception of the empty atom. Just as the planets orbit around the sun, the electrons circle around the nucleus, with its tightly packed protons and neutrons. In comparison with the nucleus, this outside band of electrons is huge: the overall size of the atom is 10,000 times that of the nucleus. "If the outer shell of electrons in the atom were the size of the Astrodome that covers the Houston baseball stadium, the nucleus would be a Ping-Pong ball in the center of the stadium. That," explains Robert Jastrow, director of the Goddard Institute for Space Studies of NASA, "is the emptiness of the atom."

?

How did ballet dancers start dancing on their toes?

The strange custom of dancing on the very tips of one's toes is not as old as ballet itself. Only after the dance was several hundred years old did this stylized position, which many now consider the essence of ballet, catch on—and one Italian girl at the Paris Opéra was largely responsible.

The classic ballet, which in the strictest sense is based on five established positions of the feet, originated in the 15th- and 16th-century Italian and Spanish courts. Here were performed "spoken dances," with group and solo dancers who acted out allegorical stories usually based on Greek myths and glorifying the King and Queen. Wandering jugglers and jesters also danced but theirs was hardly a respectable occupation. It was in France that the classic ballet became a professional art and one sanctioned by the highest nobility. Louis XIV himself danced, and the appellation Le Roi Soleil (the Sun King) derives from a ballet written for him in

An ethereal Marie Taglioni in La Sylphide, *after a painting by A. E. Chalon, R.A.*

which he appeared as the sun. Dance then moved into the theater, and the supervisor of the King's music, Jean Baptiste Lully, working with Beauchamp, Racine, and Molière, initiated the form of opera-ballet. This became so popular that in 1713 a dancing school was founded in association with the Paris Opéra.

At this time the professional female dancer began to emerge, although her role on stage was still inferior to the male's. For many years the ladies' technique remained rough, partly because they were encumbered by floor-length dresses. In the mid-18th century, virtuoso Marie Ann de Cupis de Camargo stunned audiences with the sight of her ankles. This was the first step toward accommodating costume to performance.

In 1827 an unlikely young lady from Italy, Marie Taglioni, danced at the Paris Opéra and launched a new style in the dance. At 23, Taglioni was frail, sallow, by no means a beauty. She did not ooze the feminine charms of her predecessors, like the voluptuous Maria Medina Vigano, for whom Beethoven wrote a minuet. Taglioni was ethereal, otherworldly, more spirit than flesh. She brought an unprecedented lightness to the stage. This was due not only to her physique but to her impeccable technique. As the daughter of a dance teacher and choreographer, Marie had trained under the finest teachers for many years. Such was her ease and control that she seemed to float across the stage. And to accent the image of moving heavenward, Taglioni danced on the tips of her toes. A few others had attempted this before, but Taglioni achieved it with new grace and artistry. Her style evoked a new feminine ideal, which was no longer blatantly sensual, but refined, abstract, and artificial.

Marie Taglioni danced throughout Europe for 20 years and was widely acclaimed as queen of the dance. A new word was added to the French vocabulary: *taglioniser*, meaning to dance like Taglioni. Few could get up on their toes as she did, however, because in those days the slippers provided virtually no support. They were made of silk or satin, and the toes were simply stuffed with cotton wool or silk. The first blocked shoes, enabling greater virtuosity of pointe work, were made in Italy late in the 19th century. Both men and women learned the pointe technique, but initially men refrained from using it on stage as it was considered effeminate. Increasingly, the diffident male dancer played a sec-

Marie's lofty style awes her brother, Paul, in L'Ombre.

ondary role to the airborne female, becoming little more than a *porteur* for his partner.

?

How did they start the breakfast cereal industry?

The Reverend Sylvester W. Graham preached that, rather than being born again, one's life could be salvaged by vegetarianism and bran. Living in New England in the early 19th century, the former Presbyterian preacher was an early champion of the low-fat, low-salt diet, brown bread as opposed to the socially sanctioned white, and fruits and vegetables as against beef or pork. If the Reverend Graham could only see how his flocks have multiplied a century and a half later. As it was, espousing coarse, unsifted flour, slightly stale bread, and lots of bran, he left his name to a flour and a cracker.

Graham had some notable followers in the 19th century—Thomas Edison, Amelia Bloomer, and Horace Greeley among them. Most critically, though, his influence led Mother Ellen Harmon White of the Adventist Church to found the Western Health Reform Institute at Battle Creek, Michigan, in 1866, where people with stomachaches, too much fat or too little, high blood pressure, and assorted other ailments might find physical and spiritual health. Dr. John Harvey Kellogg, acting as manager, changed the name to the Battle Creek Sanitarium and, along with his brother, developed a cereal called Granose, an immediate success. Among the patients at Battle Creek was Charles W. Post. While the absence of meat and lack of stimulants did not cure his ulcer, Post did invent Postum and Grape Nuts, the latter called Elijah's Manna until marketing problems provoked a change of name. The sanitarium must have been a creative place because there, too, Dr. Kellogg invented the corn flake for a suffering patient who broke her false teeth on a chunk of hard egg bread.

?

How did they perform the first cesarean delivery?

The cesarean section has become so quick and safe in recent years that many doctors opt for it rather than waste precious time during a prolonged labor and risk the complications that might accompany a vaginal delivery. The operation has in fact become so popular among obstetricians that despite increasing awareness among women of natural childbirth methods, cesarean deliveries far exceed the 12 percent incidence of need. Estimates range from 17 to a whopping 30 percent. In the long and remarkable history of obstetrics this is a very new development. Heated controversy surrounded the operation for many years, and the mortality rate of the child but more particularly the mother remained prohibitively high until this century.

Egyptian folklore contains numerous references to cesarean deliveries, as do some Greek and Roman texts. But in those days the operation was performed only on dead or mortally ill women, for though it was fairly simple to make an incision and remove the child, it was far more difficult to stop the hemorrhaging, prevent infection, and suture the uterus and wound. Mythological instances such as the birth of Aesculapius, snatched from Cornius's uterus as she was carried to the funeral pyre, contributed to the belief that such a birth conferred supernatural or at least heroic powers. Indeed, a popular belief persists that Julius Caesar was born in this way and gave his name to the method. But the fact that Caesar wrote to his mother during his years at war disproves this theory, since she would not have survived such a delivery. Years earlier Numa Pompilius (715–673 B.C.) had codified as law that the operation might be only performed on a woman dying in her last weeks of pregnancy. Under the emperors this *lex regia* became *lex caesarea,* hence the origin of the name.

The Talmud states as early as A.D. 400 that in the case of an abdominal delivery a woman was excused from the days of purification required after normal delivery. But the first recorded case of a successful cesarean section on a living woman who continued to live (and even bear other children) does not appear until the Renaissance. The incident was recorded 88 years after the fact by Gaspard Bauhin. This writer maintained that in 1500 in Sigerhausen, Switzerland, a swine gelder named Jacob Nufer opened the abdomen and uterus of his wife. Certainly he would have had the necessary sharp tools, and when several stone cutters and 13 midwives could do nothing to aid his wife's long and painful labor, Nufer took her fate in his own hands. Obtaining permission from the civil authorities to perform the operation, he returned to his humble house to find that all but two of the midwives had fled in terror. Undaunted, he neatly made an incision, extracted his child, and sutured the opening. The healthy child lived to the ripe old age of 77, and Nufer's wife went on to produce twins and then four more babies, all vaginally. Nufer was far ahead of his time, for it would be centuries before the operation was regularly—and safely—performed without a hysterectomy as well. Another early cesarean was performed by Christopher Bain in Italy in 1540. This time, according to Marcellus Donatus, the baby was born dead but the mother lived.

Scipione Mercurio, born in Rome in 1550, was an early champion of the frightening operation, the first to advise it for a living woman with a contracted pelvis. Early abandoning his studies in theology, Mercurio traveled and studied in France and Spain, finally settling in Venice to practice medicine. There he wrote a long and celebrated book on obstetrics, *La Commare o Riccoglitrice,* containing a lucid but rather gruesome account of the cesarean section. No anesthesia was used. The poor woman was pinned down in a large bed by five men, or hardy women, as the surgeon outlined the line of incision in ink, then cut from the pubis to just above the umbilicus. After lifting out the infant he sponged out both uterus and abdominal cavity. Pushing aside the intestines, he closed the abdomen wall with interrupted sutures. "This is enough," Mercurio abruptly concludes his long treatise, "about this new method of aiding difficult deliveries to help miserable patients."

Horror continued to surround the operation for centuries and for good reason. Although Jean Louis Baudelocque (1748–1810), one of the greatest of French obstetricians, advocated the operation when other methods of delivery failed, 42 deaths resulted from his 73 operations. He was called an assassin and was sued in at least one instance. Mortality rates in cesarean deliveries approximated 75 percent up until the mid-19th century.

The first successful cesarean sections in the United States were performed early in that century. In Edom, Virginia, Mrs. Jesse Bennett decided to sacrifice her own life for her child and acquiesced to the operation. Her husband, a doctor, assisted by a Dr. Humphrey, placed the woman on planks stretched between two barrels. Laudanum was the only anesthesia. Two servants were on hand to control Mrs. Bennett, who, miraculously, survived the procedure to live another 36 years. She was sterilized in the process, but her child lived. In 1827 John L. Richmond performed a cesarean section using a pocketknife on a black woman lying on a kitchen table. The woman recovered in 24 days and her child survived.

It may be that the cesarean method of delivery was understood and effectively carried out by illiterate peoples long before these recorded instances. Proof may never be found, and techniques may be lost forever, but in 1879 a Dr. Felkin witnessed a very skillful operation in Uganda. The mother drank a hefty dose of banana wine; then her abdomen and the specialist's hands were washed with the liquor and thereby disinfected. The operator cut through the abdominal wall and uterus, removed the child and the placenta, massaged the uterus, and elevated the patient so that the fluids would drain. The incision was closed with pins and eight sutures and covered with crushed herbs. The speed and success of the procedure have led to speculation that it had been practiced there for years, perhaps centuries.

?

How did the blue blazer become a classic?

It starts with the school uniform, insignia on the pocket, and persists for years thereafter—at dinners, offices, weddings, clubs—the blue blazer cropping up everywhere with the same safe respectability. Men of all ages turn to it with relief as an easy answer to what to wear on just about any occasion. Casual yet classic, dressy but not overly so, a cut above the same jacket in green, beige, plaid, or stripes—why this seemingly arbitrary choice?

The blue blazer originated, aptly enough, on the HMS *Blazer*, a British naval ship of the mid-19th century. One day, so legend has it, the captain of the ship despaired at the sight of his motley and tattered crew and ordered them all to wear blue serge jackets with Navy buttons. The fashion soon caught on in England's schools and clubs and crossed the ocean to find a vast market in America, where scores of retailers make thousands each year from the lasting impulse of a meticulous captain.

?

How did they decide which horses were Thoroughbreds?

This question implies a decision as arbitrary and meaningless as establishing German shepherds as Thoroughbreds, while Labradors, terriers, and the rest of the canine hoi polloi are left in a common, plebian pool. It implies, furthermore, a popular misconception of the term as meaning purebred; in the horse world the Thoroughbred is an actual, distinct (if pretentiously named) breed. Although of mixed origins, the Thoroughbred breeds true to type and has held a place alongside other breeds in the *General Stud Book* of the English Jockey Club since the 18th century. A more suitable question, then, is how did they breed the Thoroughbred, a variety of horse for which people will pay millions?

Horse racing has been popular in England since Roman times, and quite possibly before that, yet the breed of the Thoroughbred is only about 300 years old. For centuries the English used Galloways, the fast ponies of Scotland and northern England, while the Irish raced ponies called hobbies. Breeding problems arose in the 16th century when heavy importation of horses from Italy and Spain diluted the native stock. At the same time, interest in racing skyrocketed. King Charles II was particularly fond of the sport, and ordered "12 extraordinary good colts" from James D'Arcy, master of the Royal Stud, each year. Poor D'Arcy, with pressure from the King, faced a stableful of heterogeneous horses. Thus he and other breeders of the time set out to establish a stock that would breed true to type. They turned to the purest of the pure: the Arabian, carefully bred in the Mideast for centuries. Between 1660 and 1760 some 200 Mideastern horses were imported, 75 percent of them stallions, and 50 percent of them Arabians. Apparently, the initial intent was for genetic homogeneity rather than speed, since few of these beautiful horses were raced. Among the imports were three particularly magnificent animals:

the Godolphin Arabian from Tunis; the Darley Arabian, a horse of impeccable breeding brought from Aleppo in 1704; and the Byerly Turk, which allegedly carried a Colonel Byerly safely from the Battle of the Boyne in 1690, leaving his enemies in the dust. These three were bred to the Royal Mares, and it is from their offspring (one descendant of each) that all modern Thoroughbreds trace their ancestry. Descent along the female line is harder to trace since the mares were simply called Arabians, Barbs, or Turks, depending on their father's type.

The latter part of the 18th century and the beginning of the 19th saw rapid increases in the quality of the new Thoroughbred line. The horses grew slightly taller, 15 to 16½ hands becoming typical, as it is today. The graceful animals were noted for their longer stride, an advantage in racing, and increasing precocity as the fashion arose to race two-year-olds rather than more mature animals.

The passion for horse racing spread to America in the late 17th century; here, the breeding of Thoroughbreds quickly became big business—in time, the largest of its kind anywhere in the world. Over 40,000 Thoroughbred foals were registered with the Jockey Club in New York in 1982, and untold others that went unregistered may also have been foaled in this country during that year. Although racing remains their primary purpose, the horses are also bred for steeplechasing, hunting, and show jumping. Different breeding, selection, and care have led to great disparity in the quality of Thoroughbreds—the skinny trail horse you might rent for an hour may have a claim on Secretariat.

For top quality Thoroughbreds the stakes are enormous. In July 1983 Sheikh Mohammed Al-Maktoum, United Arab Emirates Defense Minister, paid $10.2 million for an offspring of the celebrated stud Nothern Dancer, the highest price ever paid for a racehorse at auction. Also eager for the yearling colt was British businessman and betting tycoon Robert Sangster. Just a year earlier Sangster had outbid the Sheikh for a son of Nijinsky II, paying $4.25 million, a record at that time.

?

How did Houdini escape from a packing case underwater?

This is "My Challenge to Death," Houdini advertised with bravura, drawing fascinated and mystified crowds to witness his spectacle of daring. They would gather in droves along a pier on a river or canal where the preeminent magician stood handcuffed, vulnerable, right by their side. Here there was no stage, no discrete and distant world where tricks and illusions were *meant* to occur.

The cold, deep water was all too real, as were the carpenters intently constructing a stout wooden crate. Houdini blithely stepped inside, his hands manacled, while the workers completed their task, hammering in wire nails and boring a hole in the bottom to make sure the case would sink. A thick rope tied around the box and firmly knotted provided the final security. Reporters and skeptics were invited to inspect the case with its precious cargo, to see and feel for themselves that there was indeed no easy way out.

As the public began to tremble, a tug, hired by Houdini, drew alongside the pier. A crane on board lifted the packing case out over the water and lowered it—to the wide-eyed disbelief of the crowds—until all trace had disappeared from sight. An excruciating few minutes ensued. Then suddenly the packing case was hoisted out of the water with Houdini, safe and sound, sitting on top, welcoming the cheers and applause of his fans. Their appreciation multiplied when, once the case was brought to the pier, all could see it was sealed, undamaged, precisely the same as it had looked on the previous inspection.

How did he do it? To begin with, the handcuffs were fake, easily opened by a secret spring, so Houdini was beyond that obstacle well before he hit the water. Second, he took with him a carefully concealed pair of nail cutters. The instant he was underwater he cut enough nails to push aside a few planks of the lid.

Houdini sets the stage for his underwater escape from a packing case.

He slipped out and sat on top of the case, pushing the boards back in place as the crane hoisted him out. Just as so many sleights of hand are effected by guiding the audience's attention away from the key action at the critical moment, so here everyone was initially so taken with the sight of Houdini, they forgot about the crate. Aboard the tug en route back to the pier, the crate was actually out of sight for a minute or two—and that was enough for two of Houdini's men to remove any broken nails and hammer in some new ones.

Despite the tools and help Houdini got with this spectacle, it was a feat of daring, requiring fast and skillful work—even for one used to holding his breath for minutes at a time, or controlling it so rigorously he could be buried alive for an hour and be resurrected. However one looks at it, the danger of being trapped underwater was considerable. ". . . It is necessary to preserve absolute serenity of spirit," remarked Houdini about his fantastic escapes. "I have to work with great delicacy and lightning speed. If I grow panicky I am lost. . . . If there is some slight accident or mishap, some slight miscalculation, I am lost unless all my faculties are free from mental tension or strain. The public . . . have no conception of the tortuous preliminary self-training that was necessary to conquer fear."

Now and then Houdini did encounter some hitches with this feat, and the British police were so sure he would drown they tried to prevent its performance. They should have known better than to try to confine the wily magician. At Liverpool, for instance, Houdini cleverly led the solicitous police down the wrong path, substituting a double for himself at the designated site. With great urgency the police arrived on the scene just as the packing case was completed. But no sooner had they stopped the show than another case 50 yards away, containing the real Houdini, was lowered into the water. Appropriately enough, the magician had misdirected the police, and the spectacle went on, to the amusement of its creator.

?

How did they discover Neanderthal man?

In 1856 some workmen were digging for limestone in a cave in the valley of the Neander River near Düsseldorf. Some old bones turned up but the workers, more interested in their paychecks than paleontology, brushed them aside and many were lost. Fortunately, the owner of the quarry salvaged a skullcap, part of a pelvic bone, a number of ribs, and some limb bones, and hustled them off to a local scientist. Johann Carl Fuhlrott confirmed they were human (although the position of this early man in human evolution was much debated in later years), and he in turn consulted with anatomist Hermann Schaaffhausen. They determined these were the bones of an ancient man—perhaps a *few* thousand years old. From the bowed limb bones and low skull with its protruding brow ridges, various anatomists concocted some inventive theories: one said the caveman had rickets, which was so painful he habitually furrowed his brow; another believed him merely to have been a Mongolian cossack who chased Napoleon across the Rhine in 1814 and retreated to the cave to die.

Many more bones of such peculiar, possibly diseased humans were unearthed throughout western Europe in the latter half of the 19th century. In 1908 Marcellin Boule of the French National Museum of History put together the first nearly complete skeleton of Neanderthal man, found in a cave in the Dordogne (the name stuck, even if far from the Neander River and its somewhat discredited discoveries). This figure was stooped and apelike; his arms dangled, his head jutted forward. Boule underscored everything bestial and crude about the Neanderthal, and until the middle of this century the image persisted of an uncouth creature with whom modern man wanted no association. But Boule failed to detect a key factor: his old man had arthritis, and this accounted for the slumped posture. Subsequent reconstruction of the skeleton by William Straus and A.J.E. Cave illustrated that the Neanderthals looked very much like us, except for the shape of the head and face. The cranium was low, the brow ridges prominent, but the skull actually held a larger brain (85 cubic inches) than ours (83 cubic inches). The Neanderthal, *Homo sapiens neanderthalensis*, averaging around five-eight in height, was a husky, muscular figure, closer to today's body builder than fashion model, primitive but definitely human. Indeed, "if he could be reincarnated," wrote Straus and Cave, "and placed in a New York subway—provided he were bathed, shaved and dressed in modern clothing—it is doubtful whether he would attract any more attention than some of its other denizens."

?

How did they build the Eiffel Tower?

On November 8, 1884, France's President Grévy signed a decree that "a Universal Exposition of Products of Industry shall be opened in Paris May 5, 1889. . . ." The last decade had seen a flourishing of trade, a vigorous expansion of industry, and rapid increases in national income. The country had salved the wounds of the Franco-Prussian War and had made an energetic comeback, expanding her colonial interests in Indochina and North and Central Africa. Thus a spirit of idealism and national pride underlay the President's decision for a world's fair, and the new Minister of Commerce and Industry, Édouard Lockroy, contributed unflagging support. Lockroy obtained a budget of $8.6 million for the exposition and enthusiastically urged the creation of a 1,000-foot tower, in the proud tradition of the pyramids and Gothic cathedrals, a symbol of France's rising powers and industrial promise, as well as man's potential to climb heavenward.

In 1886 Lockroy invited architects and engineers to submit plans for a tower or, if they wished, some other monumental structure. Among the 100 proposals received was one calling for a tower in the form of an enormous sprinkler that might rescue Paris in times of drought; another described a guillotine-shaped structure to glorify the Revolution, casting its morbid shadow over designs of despotism and oppression. But the winner had a plan very similar to the image Lockroy had been nurturing, and the credentials of the contestant who submitted it were impeccable.

Gustave Eiffel, a graduate of the École Centrale des Arts et Manufactures, was a prestigious builder in metal, noted particularly for his successful railway bridges. Earlier, as a modest young man of 22, Eiffel had planned to work at his uncle's vinegar distillery in Dijon until a family quarrel fortuitously disrupted his plans and set his sights in another direction. Eiffel landed in a Paris construction firm specializing in railway equipment, where years of experience eventually allowed him to open his own metalworking shops, obtaining contracts nationally and abroad for train stations, exhibition halls, and railway bridges. In 1886 Eiffel submitted a detailed plan for an immense 1,000-foot tower of wrought iron, a rigid but relatively light material, to be built on the Champ de Mars on a budget of $1.6 million. He cannot be given credit for the design, however, which was conceived by his chief of research, Maurice Koechlin, aided by Émile Nouguier and architect Stephen Sauvestre. The elegant tower, with its incurving edges and stupendous height, fit the judges' bill as "an original masterpiece of the metals industry" and on January 8, 1887, the contract was signed, giving Eiffel a mere two years to fulfill the task.

Not all of Paris supported the glorious monument to an age of science and industry. Writers, painters, and sculptors collaborated in righteous protestation against "the useless and monstrous Eiffel Tower, which the public has scornfully and rightfully dubbed the Tower of Babel."

The plaints proved futile and construction began. First came extensive investigation of the land on which the tower would stand. Paris lies on a deep layer of plastic, gray clay that could well support the foundations; however, near the Seine, where two of the tower's legs would rest, was a 20-foot-thick bed of unstable sand and gravel. Eiffel used pneumatic caissons to dig to a depth of 53 feet before he at last reached firm clay. The piers of the north and west supports would therefore have to be dug 16 feet deeper than those at the south and east.

The laying of the foundation alone was a massive undertaking, requiring five months and the use of 16 large iron caissons with sharp cutting edges along the bottom. With four caissons for each pier, workers removed 40,500 cubic yards of earth through air-tight hatches and successfully prevented water seepage. The caissons were illuminated with electric lights, a novel luxury at the time, and unlike Washington Augustus Roebling and his workers at the Brooklyn Bridge, no one suffered from the bends during the digging. The sturdy foundation in which each pier was set comprised 20 feet of cement, huge blocks of limestone from central France, and two layers of hard stone from Château-Landon. In each the workers set two anchor bolts 26 feet long and 4 inches in diameter, and attached a cylindrical iron shoe. A column, in turn, was bolted to each shoe. According to Joseph Harriss in his thorough book about the tower, the pressure from the columns was to be 10.2 pounds per square inch, while the foundations could accommodate over 400 pounds per square inch. The most ingenious element was Eiffel's placement of a hydraulic jack containing a piston operated by water pressure in each cylindrical shoe: when the piers rose to the point where they were connected by a horizontal platform, they could be raised or lowered minute amounts to assure that the platform was perfectly level.

Next, horse-drawn drays began to arrive at the construction site bearing wrought-iron pieces from Eiffel's shop three miles away. There, 30 draftsman had made 5,329 mechanical drawings for 18,038 separate pieces, and now the shops were pumping out 400 tons of girders and trusses each month. To facilitate handling, individual pieces were kept small, weighing no more than three tons each. Workers used steam-powered cranes to fix the pieces, and when these could no longer reach, a fantastic system of "creeper cranes" was developed. These nimble 13-ton machines, which could pivot 360°, were placed on inclined tracks, where the elevators would later run up the piers, and skillfully operated from the ground.

The four piers would not be joined until they reached a height of 180 feet, but due to their inclination at an angle of 54°, Eiffel calculated they would collapse when only half that tall. Thus, at 90 feet a temporary bracing of 12 wood pylons topped with platforms had to be erected. On these Eiffel constructed another system of hydraulic jacks. The columns leaned against metal cylinders filled with sand and each equipped with a plug and piston. If a column needed to be lowered, the plug was removed and the requisite amount of sand allowed to drain out until the correct position was attained. Raising the column could be achieved by the action of the piston. A further support system was erected above the pylons: at a height of 150 feet, four 80-foot-long platforms spanned the sides of the tower and were joined to resist the inward pressure of the piers.

Between July and October 1887, the tower climbed to 92 feet, continuing upward with incredible speed, largely due to the technique of prefabrication. Girders and trusses were already shaped and drilled so that on-site workers could simply set them in place and drive in the rivets—an average of 1,650 per day. As one newspaper reported at the time:

> . . . 250 workmen came and went in a perfectly orderly way, carrying long beams on their shoulders, climbing up and down through the lattice ironwork with surprising agility. The rapid hammer blows of the riveters could be heard, and they worked with fire that burned with the clear trembling flame of will-o'-the-wisp. The four cranes—one for each pillar—which brought up the pieces for this vast metallic framework one by one, stood against the sky with their great arms at the four corners of this lofty sight. During short winter days, when night falls the twenty rivet forges blaze in the high wind, casting a sinister glow over this tangle of girders until it acquires a fantastic aspect. The men work as late as possible and move about like shadows between these dark red smoky fires.

The workers were not always so orderly, in fact. Wages ranged from 8 cents to 14 cents an hour; many workers demanded more as the tower grew and became more dangerous, and as winter days shortened their hours. Two strikes occurred, the first on September 19, 1888, when workers demanded an immediate 4-cent-per-hour raise. Eiffel resolved the matter with a graduated raise of 1 cent per hour each month, up to a maximum raise of 4 cents an hour. When workers again put down their tools in December, Eiffel had less patience and fired anyone who failed to work. Few persisted in the strike, and the names of 199 loyal workers were immortalized on a prominent girder.

Eiffel's tour de force.

At a height of 180 feet came one of the most telling and delicate maneuvers of the entire construction. Here the four piers were joined by a 25-foot-deep iron belt. Its component parts were raised by crane to the top of the scaffolding platforms, riveted together, and carefully slid into place. It was at this point that the workers manned the various hydraulic jacks, adjusting to the finest degree the gigantic 440-ton piers. Measurements had been taken throughout construction with steel wires, and Eiffel had been careful to make the columns, if anything, too vertical so that in the fine tuning they could be lowered rather than raised. His meticulous plans paid off: in March 1888 the first platform found its place, rivet holes perfectly aligned, each pier secured with iron wedges.

Tracks were laid around this platform, along which little wagons carried material from a central crane to four creeper cranes. Eiffel installed a canteen up there to save his workers the trouble of coming down for lunch—and to monitor their alcohol intake. The second platform, at a height of 380 feet, was completed in July and fireworks were launched from it on the 14th. Just above this platform the piers converged into a single spire. Here a new system of counterbalancing cranes was employed. They ran up and down an auxiliary iron frame 30 feet high, on top of which yet another frame was mounted. With four hoisting systems operating at once, workers could get a large girder from the ground to the top and in place in a mere 20 minutes.

As the tower rose to completion Eiffel turned to the complex problem of elevators, in itself a lengthy story. Suffice it to say that because aesthetics prevented the installment of a single vertical shaft, which would have cluttered the lovely open arches of the tower, Eiffel ended up with three different elevator systems installed by three different companies. Two French firms contracted for the runs from the ground to platform 1 and from platform 2 straight up to the top. But no Frenchman dared undertake the tricky stretch between platform 1 and platform 2, where the angle of inclination shifted from 54° to 78°; this left Eiffel in a bind, for his contract prohibited the use of foreign equipment. In the end he had no choice and contracted with the American Elevator Company, a branch of Otis, for two elevators: one from the ground to platform 2, stopping at platform 1 on the north pier;

and one from platform 1 to 2 on the south pier. Otis's novel hydraulic rope-geared system was initially met with suspicion and trepidation by the Europeans, but in the end they were duly impressed. The Otis elevators carried 40 passengers 400 feet per minute, while the far noisier French "direct plunger" system (in which a piston pushed the car along hundreds of links in a chain) from the ground to platform 1 moved half as fast but accommodated 100 people. All the elevators were powered by water from the Seine, stored in reservoirs on the first and second platforms. One could make the entire trip to the top in seven minutes.

Nothing warms a Frenchman's heart like sitting down to a fine meal, so the decision about what would draw crowds to the magnificent tower was probably not a difficult one. Four restaurants opened on the first platform, including a branch of the Parisian restaurant Brébant in Louis XV decor; one could sample Anglo-American cuisine or Russian or that of Alsace-Lorraine. Platform 2 featured a printing press that issued *Le Figaro* with daily reports of exposition activities, plus a little refreshment in the way of drinks and pastry. The third platform, at the dizzying height of 905 feet, was enclosed in glass to protect visitors from the wind. And just above it was an apartment for Eiffel, for scientific experiments and astronomical and meteorological observations. The top was crowned with eight lightning rods and a powerful light that beamed 120 miles, flashing blue, white, and red. Finally, the tower was bathed in Barbados bronze, a reddish-brown paint shaded deeper near the bottom, turning gradually to gold and pale yellow as the tower rose.

Upon completion the tower was the highest structure ever made by man, standing 985 feet, 11 inches, weighing 9,441 tons, standing on a base area of 2.54 acres. Eiffel not only completed construction over a month before the exposition opened, he also succeeded in staying within his budget. Nearly 2 million visitors celebrated the exposition and the tower was toasted with speeches and tears of national pride, while Eiffel was made an officer in the Legion of Honor. Nonetheless, one spurious story circulated that Guy de Maupassant religiously lunched on the tower's second platform because it was the only place in Paris where he could not see the tower itself displaying its scar on the city's horizon.

?

How did they decide that December 25 was Christmas?

The reasons underlying the celebration of Christmas on December 25 are not entirely Christian. For one, no one really knows when Jesus was born. No one worried much about the birthday until the 3rd century, at which time Clement of Alexandria proposed May 20, but speculation continued. In 336 the Western Church designated December 25, while the Eastern Church favored January 6, which the Armenian Church maintains today.

By placing this major religious event in the winter, the Christian Church was competing with pagan celebrations that had persisted for centuries. Late December is the time of the winter solstice, when the sun is at its weakest and begins its gradual increase, marking a symbolic death and rebirth. For the Romans, December 25 was the Natalis Solis Invicti, the Birthday of the Unconquerable Sun. Threatened by the pagan cults, the newly established Church took an "if you can't beat them, join them" approach, ultimately absorbing the pagan symbolism and being strengthened by it.

?

How did Humphrey Bogart cure a hangover?

He got drunk again. And to make sure he had some company he formed his own drinking club, the Holmby Hills Rat Pack, with his wife Lauren Bacall as den mother, Frank Sinatra as secretary. Members were by and large Hollywood Democrats with an inclination for the bottle. Regulars included David Niven,

Prince Michael Romanoff, and Judy Garland, while Adlai Stevenson, John Huston, John O'Hara, Peter Lorre, and Charles Addams would show up from time to time. A featured highlight of these evenings was "killer croquet." Bogie would toss down untold drinks as he puffed through his habitual fifth pack of Chesterfields for the day.

By way of comparison, Errol Flynn conned doctors into giving him narcotics shots, usually cocaine, when he had dipped into the bottle too heavily. Richard Burton turned to chile and beer for breakfast, a menu he ordered so frequently that Elizabeth Taylor got to like it, too.

?

How did the FBI devise the "Ten Most Wanted" list?

We've all seen mug shots in post offices and police stations of shady, unshaven characters with deep, mean eyes and unkempt clothes, and most of us would prefer not to cross their paths—though apprehending them if we do is the reason for the FBI's list. Back in the '40s the Associated Press, the United Press, and the International News Service were distributing stories about major criminals wanted by the FBI. It was in 1950 that the bureau decided to collaborate with the International News Service to select and publicize a list of ten particular criminals.

Supervisors at the FBI headquarters in Washington, D.C., determine "The Ten Most Wanted Fugitives" by using the following criteria:

A. The individual must have a lengthy record of committing serious crimes and/or be considered a particularly dangerous menace to society due to current criminal charges.
B. It must be felt that the nationwide publicity afforded by the program can be of assistance in apprehending the fugitive,

who in turn should not already be notorious due to other publicity.

The list has carried 379 fugitives, including 6 women: Angela Davis, the radical leftist, was on it for two months in 1970, and Katherine Ann Power, wanted for murder, has been on it since 1970. Among the six men who made the list more than once was James Earl Ray, assassin of Martin Luther King, Jr., posted in 1968 and again in 1977. Three hundred fifty-three of the fugitives have been apprehended, nearly a third as a result of citizen cooperation. (Detractors of former FBI director J. Edgar Hoover maintained that Hoover only placed on the list criminals who were on the brink of capture—in the process enhancing the bureau's public image.) From these 353 the FBI has come up with an average profile of a Top Tenner: height, 5 feet, 9 inches; weight, 167 pounds; age at apprehension, 36; average time on list, 157 days; average distance between crime scene and apprehension point, 969 miles. Charles Lee Herron, still at large and wanted for murder, holds the record for tenure on the list: 14 years of skillful evasion of the FBI.

?

How did they develop the Salk vaccine?

An Egyptian stele from about 1500 B.C. portrays a young man with a withered leg, symptomatic of poliomyelitis. *The Procession of the Cripples* by Hieronymus Bosch also captures the cruel effects of the disease. In the late 19th century and first half of the 20th, what had been isolated instances erupted, unpredictably, into devastating epidemics. Outbreaks in Sweden in 1881, 1887, and 1905 killed hundreds of children and crippled over 1,000. The year 1916 saw a severe epidemic in the United States with over 27,000 cases and over 7,000 deaths in the 20 states requiring victims to file a report. New York City, where over 9,000 were hit, was panic-stricken. Many bolted their doors and locked their windows in terror of infection. Others fled to the suburbs and effectively spread the disease. At this time it was believed that polio entered through the nose and passed on to the brain. Some tried zinc-plating their nostrils.

One of the first breakthroughs in tracking the elusive disease came in 1909. Medical scientist Karl Landsteiner of Vienna removed nerve cells from the spinal cord of a four-year-old killed by polio. He injected these cells into a baboon, which subsequently died, and a monkey, which became crippled. He kept the infectious microbe alive in a broth made from nerve tissue and went on to find that one monkey, able to survive a first infection, was thereafter immune altogether: something in its blood serum, in fact, killed the microbes in the test tube. In the same year, Simon Flexner and P. A. Lewis of the Rockefeller Institute succeeded in transmitting a human infection from one monkey to another. The infectious agent, which could multiply within its victim, was able to pass through extremely fine filters and could not be seen. (That would change with the discovery of the electron microscope in 1931.) Flexner and Lewis called the agent a filterable virus. Despite the now obvious clue in their findings, too little was known at the time about viral infection and immunization for anyone to bring the evidence to a fruitful conclusion. Polio continued to find its victims: not only infants but, increasingly, older children and adults. Franklin D. Roosevelt was stricken at 40.

By 1935 little had been revealed about the nature of the polio virus. Nevertheless, two men independently created vaccines and prematurely injected thousands of children—with disastrous results. One was Maurice Brodie, who, with Dr. William H. Park of the New York City Department of Health, ground up spinal cords of monkeys infected with polio, then inactivated the virus with formaldehyde. After injecting only 20 monkeys with the solution, he tried it on himself and 3,000 children. Several got polio and one died. Dr. John A. Kolmer of Temple University in Philadelphia concocted a vaccine employing a live, but chemically altered, virus. His lethal potion resulted in 6 deaths among the 12,000 injected. These cases naturally left the public somewhat skeptical about vaccines and did nothing to support the use of killed virus, held in disregard by the scientific community, but championed, finally, by one Jonas E. Salk.

Salk was an extremely bright, ambitious, inscrutable young man who grew up at 106th Street and Madison Avenue in New York City and took his medical degree from New York University, although he did not want to be a doctor. Salk's interests lay in the challenging and largely unexplored field of virology. As a research fellow at the University of Michigan he developed a vaccine against the influenza virus. In 1947 he became director of the Virus Research Laboratory of the University of Pittsburgh's School of Medicine, and the following year was approached by the National Foundation of Infant Paralysis.

Established by FDR in 1938, the foundation was spearheaded by a flamboyant and celebrated Wall Street lawyer, Basil O'Connor. When the foundation's scientists were delayed in their polio research by a dearth of monkeys, O'Connor launched a massive monkey business, gaining export clearances from India and the Philippines. Over 17,000 were used in a virus-typing program in which Salk took part. From the start, Salk concentrated on the immunological rather than the infectious capacity of viruses and furthermore believed in the immunological effectiveness of killed virus. On both stands he faced open hostility in the established scientific community.

In 1949—a year in which 43,000 cases of polio were reported in the United States—a revolutionary discovery was made by Boston scientists John F. Enders, Thomas H. Weller, and Frederick C. Robbins, who later received a Nobel Prize for their efforts. The Boston group succeeded in growing polio virus in a test tube of nonnervous human tissue. Since the virus could thrive in cells other than nerve cells, its entry point might not be the nerves of the nose. Subsequent experiments over the next few years would show that the microbes were swallowed and passed through the intestine, where they multiplied and entered the circulation, thereby reaching the central nervous system and damaging muscle controls situated at the base of the brain. This circuitous route was welcome news. Now scientists had some hope of intercepting the lethal virus before it struck the nervous tissue.

Although the foundation originally thought Salk should confine his efforts to the virus-typing program, he forged ahead, convinced that polio could be prevented by immunological means.

He obtained tissue cultures from Enders, and in 1950 a grant from the foundation. Aided by Julius Youngner and Elsie Ward, Salk studied the effects of different strains of virus on tissue cultures. "Three of them," reported Salk, "gave brilliant, startling results, destroying monkey and human tissue right before our eyes. It was thrilling. These three strains, chosen at random to see how they might behave in the test tube, were obviously the best candidates for our subsequent experiments with mice and monkeys. Lo and behold, they then turned out to be the strains best suited for the experimental vaccines we later tested in human beings. They were the most antigenic strains, the most stable, the most reliable. It was a fantastic accident, choosing those particular strains as we did. Others have spent years trying to find better ones. So have I. But nobody has found one."

Salk soon began immunizing monkeys with his viruses, attenuated by exposure to Formalin. But the number of variables and details was still staggering: what amount of Formalin was necessary to inactivate the virus but not destroy its capacity to serve as an antigen? Salk had to determine proportions that could be used with certainty by future generations. He had to be sure no fraction of the virus remained active—and potentially lethal. Salk studied the chemical reactions between poliovirus and Formalin and evolved a formula for the inactivation process: virus and Formalin combined in a proportion of 250 to 1 at a temperature of 33–34°F for one to three weeks. Adjustments were constantly made in the nonnervous monkey tissue to obtain cultures most nourishing to the viral strains. Kidney tissue (rather than testes, which had been used previously) proved most effective, and the Connaught Laboratories of the University of Toronto soon produced a synthetic nutrient solution to sustain these tissues. By 1952, there was still much to do. ". . . All we had," said Salk, "were several dozen different experimental preparations, some with adjuvant [oil emulsions that increase the antigenicity of viruses], some without, some containing one type of virus, some another or a third or all three, some made with monkey kidney, some with testes, some inactivated for ten days, some for thirteen, some for twenty-one." Salk continued to work meticulously, indefatigably, and soon took the courageous step of endorsing human experimentation.

In June 1952 Jonas Salk obtained blood samples from 45 children and 27 members of the staff of the D. T. Watson Home for Crippled Children in Leetsdale, Pennsylvania. He typed the antibodies in the blood in order to determine which type of vaccine to use. Salk knew there were three types of poliovirus and that each would produce immunity only to the same type. (Thus one person could have three attacks unless inoculated against all three strains.) On July 2, he gave 43 children either Type I, II, or III vaccine. "When you inoculate children with a polio vaccine," commented Salk, "you don't sleep well for two or three months." Later, he also injected people who had no antibodies at all. He waited in nervous anticipation throughout the summer. When he again took blood samples he found that those which had had some antibodies to begin with now had far more, and those which had begun with none now showed high levels. He placed these blood samples in tissue cultures containing poliovirus. Miraculously, the *tissue* cells continued to grow. The antibodies had done their work on the virulent virus.

On March 26, 1953, Salk spoke to the nation on the CBS radio network and overnight became a hero in the eyes of a public long plagued and bewildered by the killing disease. He outlined his experiments and findings to date. Antibodies effective against the three types of virus had been stimulated in 160 persons without ill effect, and at levels comparable to those produced after a natural infection. While the nation eagerly waited, Salk stressed the need for further experimentation before a dependable vaccine could be made available. So throughout the year he continued to work. Nineteen fifty-four saw a massive field trial in which 1,830,000 children were inoculated, with positive results. In April 1955, news broke that Jonas Salk had developed a safe and effective vaccination. The U.S. Department of Health, Education, and Welfare licensed its commercial production, and plans were made to inoculate 30 million children that year.

Americans, of whom 100 million had supported the venture through the March of Dimes and another 7 million had collaborated in some capacity with the foundation, now longed to embrace their hero. They sent money, flowers, food, telegrams; they offered cars and glamorous vacations; colleges offered honorary degrees; and the press went wild. Salk did not welcome such adulation and fame. It disrupted his work and embarrassed him. Then only 40, he simply wanted to get back to the lab.

?

How did they name the Edsel?

After Edsel Ford, the son of Henry, of course—but only after a long, convoluted, and expensive chain of experiments, arguments, marketing research, and trial and error. And after all that, the adventuresome car for the upwardly mobile family was branded with an old-fashioned, grandfatherly name, scarcely well suited for a stylish new car, with its horse-collar-shaped radiator grille, wide wings on the rear, and steering wheel bedecked with push buttons—epitomizing right there the spirit of a new era.

The 1958 Edsel, a car for the prosperous that lost millions.

While the initial idea for a brand-new medium-priced car dates to 1948, when Henry Ford II suggested it to Ernest R. Breech, executive vice-president, a firm decision to launch such a car was not made until 1955. And in that year car manufacturers had no cause for trepidation. The American public, now fully recovered from the Korean War, discovered unprecedented prosperity and the big car became an outward symbol for that affluence. Chevrolets and Plymouths were getting bigger and more varied. They were being equipped with power steering, power brakes, push-button-controlled windows, and the competition was on for increasingly higher engine power. Everyone wanted a car and a buying record—7 million units—was set that year.

As head of the special products division at Ford, R. E. Krafve oversaw the birth of the new car. For a long time it was referred to as the E-Car, *E* standing for experimental, though quite early on Krafve suggested it be called the Edsel. Edsel's three sons—Henry II, Benson, and William Clay—thought otherwise, however, rejecting out of hand the blatant commercialization of their late father's name. Little more was said as Roy A. Brown, put in charge of design, went to work in celebrated secrecy, with security guards at the doors 24 hours a day. Brown had no guiding flashes of intuition, but plodded along, part by part.

Meanwhile, David Wallace, director of planning for market research, set out to determine just what image the public was after and and just what name would encapsulate it. ". . . Cars are the means to a sort of dream fulfillment," said Wallace. "There's some irrational factor in people that makes them want one kind of car rather than another—something that has nothing to do with the mechanism at all but with the car's personality, as the customer imagines it." What was to be the personality of the new car? From interviews of 800 recent car buyers in Peoria, Illinois, and 800 more in San Bernardino, California, Wallace learned that Ford's only medium-priced car, the Mercury, was considered virtually a hot rod, something young, fast, and male—and for the low-income bracket, though in reality it was not. Thus as people moved to higher incomes, they turned to a non-Ford car for something more respectable. The E-Car had to catch the eye of people on the rise, not too old or too young, and not exclusively male, though the car's appearance should in no way be feminine. Wallace outlined the desired image as follows:

> The most advantageous personality for the E-Car might well be THE SMART CAR FOR THE YOUNGER EXECUTIVE OR PROFESSIONAL FAMILY ON ITS WAY UP.
>
> Smart car: recognition by others of the owner's good style and taste.
>
> Younger: appealing to the spirited but responsible adventurers.
>
> Executive or professional: millions pretend to this status, whether they can attain it or not.
>
> Family: not exclusively masculine; a wholesome "good" role.
>
> On Its Way Up: "The E-Car has faith in you, son; we'll help you make it!"

Wallace then sent researchers out onto the streets of Chicago, New York, and Ann Arbor and Ypsilanti, Michigan. Armed with 2,000 potential names for the car, the interviewers asked people not only what they thought of the name, but what free associations each evoked, and what they considered the opposite to be. The research accumulated masses of material, but nothing conclusive.

Day after day, Krafve and other Ford executives sat in a dark room reacting to flash cards with such suggested names as Altair, Ariel, Dart, Mars, Jupiter, and Phoenix. But no agreement could be reached.

There followed one of the most unusual twists in the history of commercial marketing. The wife of one of Ford's junior assistants had recently graduated from Mount Holyoke College, where she had heard and been duly impressed by the poet Marianne Moore. The young graduate suggested that Wallace ask Miss Moore what *she* would call the car and Wallace took her up on it. "We should like this name," wrote Wallace to the distinguished poet, ". . . to convey, through association or other conjuration, some visceral feeling of elegance, fleetness, advanced features and design." Moore responded with such lofty titles as Intelligent Bullet, Andante con Moto, Utopian Turtletop, Pastelogram, and Bullet Cloisonné. Somehow, Ford didn't bite.

Next, the prominent Madison Avenue advertising agency Foote, Cone, and Belding was hired to solve the problem. They conducted a competition in their New York, London, and Chicago offices to name the car, inciting their employees with the prospect of winning a car. The agency soon had a dizzying list of 18,000 names, which they cut by two thirds before presenting it to Ford. But the irate executives at Ford did not wish to cope with 6,000—they wanted only 1—and asked Foote, Cone to cut the list to 10. During a crash program over one weekend, the New York and Chicago offices of Foote, Cone independently trimmed the list and amazingly enough came up with four of the same names on their lists of ten: Corsair was the leading contender, followed by Citation, Pacer, and Ranger. Wallace liked Corsair, too, and noted that it had rated high on the street interviews, sparking such free-spirited associations as "pirate" and "swashbuckler."

The Ford executive committee met to make the final choice. All three brothers were away. Ernest Breech took one look at the top ten and announced he didn't like any of them. The group turned to some of the past rejects and the brusque, impatient Breech lighted on Edsel. That was the one he wanted and he got a majority to agree, with Corsair and the other three favorites on hand as subnames for variations of the E-Car. Breech called Henry II down in Nassau and obtained his approval of the choice, and he in turn got his brothers to concede.

Whether because of its name, the recession of 1958, the crowded market of middle-priced cars, the time gap between the car's design and its presentation to the public, or its "lemon-sucking" grille, the Edsel was a disaster. It was launched in 1957 to the accompaniment of unprecedented advertising and publicity campaigns. Two years later only 100,000 cars had been sold. Ford discontinued the line in November 1959 having lost, reported John Brooks in *The New Yorker*, a staggering $350 million.

?

How did they choose Vivien Leigh to play Scarlett O'Hara in *Gone With the Wind*?

When the shooting of *Gone With the Wind* began, producer David O. Selznick was still searching for Scarlett O'Hara. Actresses flocked to Hollywood, talent scouts combed the South, and Selznick's backers banged on his door. Over the course of two and a half years he'd shot 165,000 feet of film (27 hours), spending $105,000 on the most extensive screen tests in the history of cinema.

While Bette Davis, Joan Fontaine, Susan Hayward, Lana Turner, Katharine Hepburn, Norma Shearer, and other luminaries vied for the role, a virtually unknown Vivien Leigh vacationed on the Riviera with Laurence Olivier and a dog-eared copy of *Gone With the Wind*. She was passionate about the book and about Scarlett and, against overwhelming odds, determined to become the lovely southerner with fiery green eyes, "a Cheshire cat smile," and 16-inch waist. She sailed to New York and flew to Los Angeles—a 15½ hour flight during which she practiced the catlike expressions of Scarlett O'Hara.

At the Selznick studio, Vivien Leigh was introduced to David by his brother Myron, who was, fortuitously, Olivier's agent. She wore a graceful black dress, tighly cinched at the waist, and a broad-brimmed black hat. ". . . The flames [of burning Atlanta] were lighting up her face," Selznick later recalled. "I took one look and knew she was right. . . . I'll never recover from that first look."

Looks aside, Vivien had to act. Final tests were made of three strong contenders—Joan Bennett; Jean Arthur; the favorite, Paulette Goddard—and of the newcomer, Vivien Leigh. These were scenes of Scarlett lacing up her corsets, confronting Ashley at Twelve Oaks, and entertaining with drunken passion Rhett Butler's proposal of marriage. When the individual scenes with different actresses were viewed back to back, Selznick was at long last able to make a decision. He'd found that actress who concealed wildness beneath elegance and style, whose "green eyes in the carefully sweet face," as Margaret Mitchell wrote, were "turbulent, lusty for life, distinctly at variance with her decorous manner."

There were only two problems with Vivien Leigh: she was English (which patriotic Americans might not find appropriate), and she was having an affair with Olivier, although both were married and had children. In his long press release announcing his choice, Selznick diplomatically avoided saying Vivien was English; she just happened to be married to a London barrister and had worked in England recently. He then visited the amorous couple and strongly urged the utmost discretion on their part. Poor Paulette Goddard had apparently been on the brink of success (before Vivien appeared), when controversey over her own ambiguous relationship with Charlie Chaplin delayed the casting. The film-going public, argued Selznick, would not look kindly on such illicit behavior. As it was, the Olivier-Leigh romance probably added to the allure of the new actress, who would charm and dazzle audiences for decades to come.

?

How did they build Central Park?

When in 1857 Frederick Law Olmsted became superintendent of the area that would become Central Park, he described it as "a pestilential spot, where rank vegetation and miasmatic odors taint every breath of air." The land was a chaotic mix of swamp and brambles, squatters' shacks and open sewers. Some 300 hovels dotted the unpromising landscape, along with hog farms, "swill-mills," and bone-boiling works. Olmsted's task was to clear the area. The intractable squatters forced him to call in the police that year, but thereafter his faithful team of 1,000 wod steadily, draining swamps, blasting rock, and carting away rubble in horse-drawn wagons.

The idea for a park had originated over a decade before. The poet William Cullen Bryant, editor of the *New York Evening Post,* was an early proponent, as was the prominent landscape architect Andrew Jackson Downing. "A large public park," wrote Downing in his magazine, *The Horticulturist,* ". . . would not only *pay* in money, but largely civilize and refine the national character, foster the love of rural beauty, and increase the knowledge of, and taste for, rare and beautiful trees and plants. . . . The true policy of republics is to foster the taste for great public libraries, parks and gardens which *all* may enjoy."

The alarming rate of urban growth also brought many to a new awareness of the beauty rather than mere utility of the natural world. Nature became a work of art, to be appreciated as such. Thus the seeds of the Romantic movement took hold, with Bryant advising us to listen to Nature, painters Thomas Cole and Asher B. Durand and many others setting up their easels in the wilds of the Catskills and the Hudson Valley. Americans were not, however, revolutionary in this. Preference for the natural as opposed to the classical, formal style of the gardens of Versailles, for example, had taken root a century before in Europe. Two branches of this revolution grew rapidly in England. One group believed that landscapes should be "improved"—that is, their beauties enhanced without destroying the curving lines and asymmetrical shapes that lent a natural aspect. Lancelot "Capability" Brown, who frequently was heard to say, "My Lord, your property has great capability," was the outstanding leader of this group—so popular that he turned down work in Ireland because he "had not yet finished England." Brown was responsible for the grounds of Sir Winston Churchill's family home. A more hysterical group were the founders of the picturesque, children of Jean Jacques Rousseau, admirers of the painters Claude Lorrain and Salvator Rosa. For them Brown's placid ponds and wandering streams were too tame. Their souls yearned for craggy peaks and gushing rivers, tangled vines and virgin forests. Both movements found their way to America and into the masterful plan for Central Park.

After rejecting an initial proposal for a park along the East River, the state legislature authorized the city in 1853 to buy the area from 59th to 106th Street between 5th and 8th Avenue. This block of 624 acres was extended to 110th Street in 1859, making a total of 843 acres, purchased for $5 million. A Board of Commissioners was established in 1857, and they decided to hold a competition for the best plan for the new park. At this point Calvert Vaux, an English-born architect and former partner of the late Andrew Downing, approached Frederick Olmsted. Landscape architecture was the farthest thing from the superintendent's mind. His scattered background included farming on Staten Island, reporting on agriculture in England, and a financially disastrous venture in publishing. Vaux, however, believed in Olmsted's artistic eye. Olmsted, perhaps less sure, was short of cash and lured by the prospect of the $2,000 prize. The two worked prodigiously, examining every inch of the park, formulating a plan to highlight its natural features—outcroppings of rock, hillsides, and areas suitable for bodies of water. Their plan, titled Greensward (it may be seen today in the Arsenal in Central Park), was an easy victor over 32 other plans for fussy gardens and sentimentally patriotic motifs.

The Greensward Plan divided the area thematically at the 85th Street Transverse Road. The upper portion, with more interesting natural features, would be left rugged, wild, picturesque. Any visible interference, by roads, buildings, or formal planting, was to be avoided. The lower portion, which was more heterogeneous, would make use of the hillside facing what is now the old reservoir. They placed the Lake south of it, with a terrace for leisurely strolls facing the rising hillside. Here a more formal element was appropriate: the straight Mall offering an elegant promenade beneath elms that led to the classical Bethesda Fountain. Farther south the land was extremely flat, and this became a meadow, dotted with grazing sheep. One of the most significant battles was with the Croton Aqueduct Board, which had plans for a new *rectangular* reservoir. Fortunately, the board yielded to Vaux and Olmsted's preference for a natural shape—the one we see today. The two planners also had to battle for their most ingenious element: four sunken transverse roads, including the first traffic underpass in America. Their plan not only would minimize the visible intrusion of traffic, but also separate the park's various activities. Pedestrians could amble over cast-iron bridges while stately carriages proceeded along the winding roads. "Winding" must be underscored, for Olmsted was determined to avoid a straightaway, which, he felt, would only invite "trotting matches."

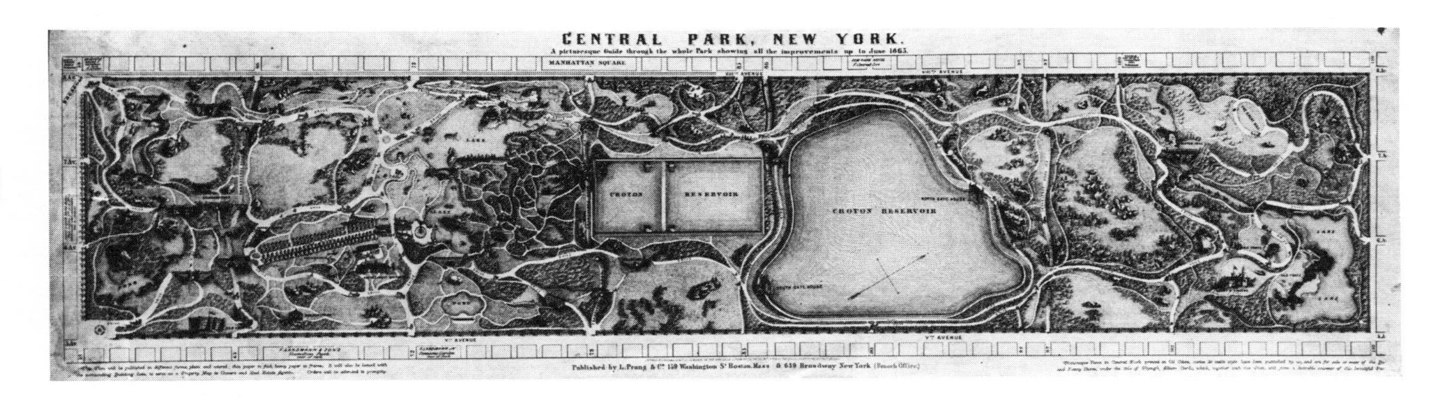

The development of Central Park as of June 1865.

In addition to his job as superintendent, Olmsted became architect in chief in 1858, with Vaux as his assistant. Together they watched their plan become a reality. The lengthy task of clearing the park continued. Blasts were heard daily from the area's swamps and crags as men dynamited the hard schist to excavate sunken roads and valleys for water. Henry Hope Reed reports in *Central Park* that over the next 15 years nearly 5 million cubic yards of stone and soil were removed, an amount requiring 10 million one-horse cartloads. And between 1858 and 1865, over 500,000 cubic yards of fertilizer were spread over the park, as were immense quantities of topsoil to enrich the area's glacial outwash, which was too poor in nutrients to support the vast array of intended vegetation.

An Austrian immigrant named Ignaz Anton Pilat, former gardener to Prince Metternich, deserves credit for the wonderful variety and naturalistic style of vegetation once seen in Central Park. In its early days one could see clear across the park, making it seem rather narrow, but flourishing trees soon dispelled this impression. By 1873, 4 million to 5 million trees, vines, and shrubs had been planted, the number of tree species increasing from 42 to 632. Evergreens were plentiful to keep the park green throughout the winter, particularly along the West Drive from 77th to 100th Street; over 800 species of perennials and alpines were introduced as well. The oldest commercial nursery in America, the William Prince Nursery in Flushing, supplied Pilat with seedlings of native and imported European and Asian species. The Parsons Nursery in Flushing also contributed, and two large nurseries were developed within the park. In addition to maples, oaks, and willows, one found Osage orange trees, Chinese elms, ginkgos, bald cypress, and golden larch. Hydrants and 12½ miles of earthen water pipes provided irrigation.

Despite this extensive work, the natural features of the land were preserved roughly in accordance with the Greensward concept. "The hips and elbows and other bones of Nature," wrote Oliver Wendell Holmes in 1862, "stick out here and there in the shape of rocks which give character to the scenery, and an unchangeable, unpurchasable look to a landscape that without them would have been in danger of being fattened by art and money out of all its native features. The roads were fine, the sheets of water beautiful, the bridges handsome, the swans elegant in their deportment, the grass green and as short as a fast horse's winter coat."

Olmsted and Vaux would shudder if they could see today's explosion of recreational facilities in the park—playgrounds, tennis courts, the Children's Zoo—which have disrupted their pastoral idyll. Their plan allowed only for quiet strolls and ice skating on the various ponds or lakes. The latter became extremely popular, with more people visiting the park in January than in July. The 20th century has brought, too, an unfortunate jumble of statues, most of questionable aesthetic value. The first was a lifeless bust of Johann Christoph Friedrich von Schiller in 1859, followed somewhat later by a group around the Mall area, including Samuel F. B. Morse, Sir Walter Scott, and William Shakespeare. A mania for statuary swept the country in the 19th century and well-intentioned donors began cluttering the park, despite the restrictions of the park's committee on statues.

Other encroachment has occurred over the years in the form of buildings. Originally there was only the Blockhouse (1814), the Arsenal (1848), which housed the Museum of Natural History from 1869 to 1877, and the two Gate Houses of the Reservoir (1863). Vaux designed the rugged Belvedere Castle, built in 1871, as well as the Ball Players' House, the Dairy, the 22nd Precinct Station (then a stable), and the Workshops, all constructed around that time. But gradually the designers realized the danger of too much building, as the Metropolitan Museum and the zoo ate up precious ground.

Olmsted and Vaux faced obstacles on the political front with the ascendancy of William Marcy Tweed and the Tammany Hall politicians. Boss Tweed created a Department of Public Works in 1870 with himself as head, and one of his men became chief of the Department of Parks, which replaced the Central Park Board of Commissioners. The park became an arena of contract letting, patronage, and graft, with the politicians' eyes on favors and kickbacks rather than the best interests of the park and its original design. The result was a disruption of planting, thoughtless manicuring of wild areas, and myriad plans for useless buildings. Olmsted and Vaux were forced to resign that year but returned in 1871 when Tweed was ousted. They continued work despite heavy financial restraints and debts accumulated in their absence. The park commission remained fundamentally under the thumb of Tammany Hall, however, and Olmsted's task of fulfilling his vision became virtually impossible. In 1878 his job was abolished and he was relegated to the position of consulting architect, with little power. Disillusioned and disenchanted with the pervasive political corruption, he soon moved to Massachusetts. Vaux continued as landscape architect to the Department of Public Works intermittently until 1895. His successor, Samuel B. Parsons, Jr., born to a love of trees as son of the nurseryman in Flushing, adhered to the ideals of the two master planners, bringing the park to a pinnacle of beauty in 1911. But his retirement in that year opened the way for further disruption of the picturesque, the disappearance of exotic trees, and inadequate maintenance, marking the end of the "Greensward Dynasty."

?

How did Peary reach the North Pole?

Admiral Robert E. Peary's ship, the *Roosevelt*, steamed north through the ice-packed waters of Smith Sound in 1908, her wooden hull specially fortified to meet the harsh barriers of ice, her primary means of propulsion a 1,000-horsepower engine rather than the sails of previous arctic exploration ships. Peary's methods differed in another fundamental way: years of experience in the north sledging across the great central plateau of Greenland had taught him the necessity of living like the Eskimos, sleeping in igloos on musk ox skins, keeping to a monotonous diet of dried meat and fat. Over and above these practical tactics integral to his success was Peary's obsession with a dream, his concept of himself solely as an instrument to attain it, and his unfaltering will—for "the grim guardians of earth's remotest spot," he wrote, "will accept no man as guest until he has been tried and tested by the severest ordeal."

As the impracticality of northern trading routes to the Far East became apparent in the 19th century, explorers nonetheless persisted in venturing north, stirred by the new goal of reaching the top of the world. Sir Edward Parry was one such adventurer who in 1827 confronted the northern regions with small iron-bottomed boats, which were sailed in open water, then dragged over the solid ice. Southern currents overwhelmed his boats, however, preventing him from getting farther north than 82°45′ N, 435 miles from the Pole. Fridtjof Nansen concocted the novel scheme of sailing a boat across the top of Siberia, freezing her into the ice pack, and simply drifting on toward the Pole. Thus the hardy *Fram,* constructed to withstand pressure from the ice, drifted from 1893 to 1895. At 83° N, Nansen and a companion left the ship to race the remaining distance in dog sleds. They got within 224 miles of their goal before drifting ice forced their return. While Nansen fortuitously hooked up with a British expedition, the *Fram* remained locked in ice for 35 grueling months, only returning to Norway in 1896. Still strong, she would sail again in 1911 under the command of Roald Amundsen, the first to reach the South Pole.

Peary himself made several attempts to attain the North Pole. In 1898, before the days of the *Roosevelt,* Peary's ship was blocked from entering the Arctic Ocean by solid ice packs. Forced to set up base camp 700 miles from the Pole, he managed to sledge half that distance before turning back. On another expedition, in 1905, violent storms and food shortages put an end to his progress at the position of 87°06′ N.

At 50, Robert Peary was scarcely ready for the comforts of home. Too much civilization made him restless. Hard as it is to fathom, he opted for the desolation of the arctic, the haze of wind-driven snow, gleaming black channels of water, and the uncanny sight of sun and moon circling the sky together. Wiser and yet more determined after his first two expeditions, Peary sailed up Smith Sound from the northwest coast of Greenland in 1908 with 49 Eskimos and 246 dogs. The *Roosevelt* steamed 350 miles north to Cape Sheridan on Ellesmere Island, crashing through "ice of all shapes and sizes," wrote Peary, "mountainous ice, flat ice, ragged and tortured ice, ice that, for every foot of height revealed above the surface of the water, hides seven feet below—

a theatre of action which for diabolic and Titanic struggle makes Dante's frozen circle of the Inferno seem like a skating pond."

Throughout the winter months Peary and his men transported supplies in 13-foot oak sleds equipped with ash runners 90 miles overland to Cape Columbia. Rations were calculated to the finest detail, allowing each man daily one pound of pemmican (dried meat with fat), one pound of ship's biscuits, four ounces of condensed milk, and one-half ounce of tea, and for the dogs, one pound of pemmican each daily.

On February 28, 1909, an advance party led by Captain Robert Bartlett set out from Cape Columbia on the northern shores of Ellesmere Island, 413 miles from the Pole. Peary's scheme was that advance parties would open the track and supply provisions part of the way, the final stretch to be tackled by his assault team in a strenuous dash. By March 1, 24 men, 19 sleds, and 133 dogs were on their way north. Leads, or channels of water, presented the greatest trial, sometimes delaying them for days, while the ice cracked, shifted, and ground together. Ice floes served as precarious stepping-stones. Once the ice shifted as they slept and Peary awoke to find a widening lead a mere foot from a team of dogs. Sleeping under skins, the men were able to get out more quickly than from sleeping bags.

The last supporting team turned back on March 30, leaving Peary with his longtime assistant Matthew Henson, 4 Eskimos, 5 sleds, 40 dogs, and 133 miles to go. With the best dogs and a lightened load, they sped over the ice, covering 25 miles in a ten-hour day for five consecutive days. Harsh winds and frigid temperatures of −35° to −40° F cracked and burned their faces. Even the Eskimos complained about their noses. In places the hazardous ice was so thin they had to crawl. On the second day a runner of one sled cut clear through the ice and was nearly lost. At about noon (of the Columbia meridian) on April 6, Peary stood at 89°57′ N, a mere three miles from his dream, and too tired to take another step. When after a few hours of sleep they resumed the journey, a peculiar phenomenon occurred. They traveled due north, then found themselves heading due south, without having changed their course. Turning back on their path, they came to that point where north, east, and west cease to exist. "Where we were," recorded Peary, "one day and one night constituted a year, a hundred such days constituted a century. Had we stood in that spot during the six months of the Arctic winter night, we should have seen every star of the Northern Hemisphere circling the sky at the same distance from the horizon, with Polaris (the North Star) practically in the zenith."

Here on this illustrious spot Peary not only planted a silk American flag, made by Mrs. Peary 15 years before, but felt compelled to leave the colors of the Delta Kappa Epsilon fraternity (with fond memories of Bowdoin College), as well as the "World's Ensign of Liberty and Peace," and flags of the Navy League and the Red Cross. One can only wonder what passed through Marie Peary's mind when she received this from her incorrigible husband:

> 90 NORTH LATITUDE, April 7th
>
> MY DEAR JO,
>
> I have won out at last. Have been here a day. I start for home and you in an hour. Love to the "kidsies."
>
> Bert

Thus 20 years of planning culminated in 30 hours at the Pole, whereupon Peary, with one backward glance, headed south.

?

How did they name the lollipop?

Apparently George Smith, who first made a candy on a stick for a Connecticut candy manufacturer in the early 1900s, had other moneymaking interests: Lolly Pop was one of the finest racehorses of the time.

?

How did they build the Mother Ship in *Close Encounters*?

During the summer of 1977 many Americans walked the streets with a distracted look in their eyes and glanced frequently, apprehensively, at the sky. That was the summer aliens hovered overhead in a chandelier spaceship, sounding a sacred chord for mystified mortals. More spectacular than an oil refinery seen from the air at night (that was the requirement the creators of the Mother Ship had to meet), the luminous UFO dazzled Roy Neary (Richard Dreyfuss) and the geniuses of science and us as well in the climax of *Close Encounters of the Third Kind*.

The real thing, if somewhat anticlimactic, is nevertheless impressive. The Mother Ship was an oval electrical system, a mere six feet in diameter. Assembled with great precision, it was made of steel, Plexiglas, plywood, fiber glass, and tiny aluminum tubes with high-voltage wiring, the whole thing surrounded by neon tubes. How did it open? For this director Steven Spielberg and special effects wizard Douglas Trumbull used another model, a ship with an eight-foot-diameter dome on which lights were projected. The final scene was shot in an airplane hangar in Mobile, Alabama. There a full-scale section of the ship, constructed of steel, was encircled by 2,000 floodlights and 6 arc lamps. Much of the body of the ship was added later in the lab. Black velvet covered the walls of the hangar, and the image of the horizon and sky was produced by front projection on a 100-by-38-foot screen. This is a technique, developed in the 1960s, that enables one to project background scenery from the front. Basically, the background image is bounced off a two-way mirror, placed at an angle to a specially treated reflex screen behind the actors. The result is an exceptionally strong, sharp image.

The Mother Ship remains a singular effect in futuristic films, so unlike the more tangible spaceships *Enterprise*, *Millennium Falcon*, and the Rebel Star Cruiser. Indeed, Trumbull compared filming his awesome UFO to "photographing God—people have a very abstract, mind's eye view of what they expect to see in a flying saucer. So the general look we went for," he explained, "was one of motion, velocity, luminosity, and brilliance. We used very sophisticated fiber optics and light-scanning techniques to modulate, control, and color light on film to create the appearance of shape when in fact no shape existed."

?

How did they discover DNA?

DNA is virtually synonymous with James Watson and Francis Crick in popular belief, largely because their contribution to its study was momentous and relatively recent. But Watson and Crick were not the discoverers of the miraculous substance that embodies the secret code of genetic characteristics. These two, who fathomed the complex *structure* of DNA, were at the culmination of nearly a century of slow, painstaking work to unravel the mysteries of the living cell and its capabilities.

The actual discovery of the substance occurred as early as 1869, and like that of penicillin, it happened accidentally. Friedrich Miescher, then only 25, was a serious, obsessively diligent scientist just launching his career. Having recently graduated from medical school in Basel, Switzerland, young Miescher cast about for an appropriate field. A hearing difficulty precluded many options, so on the advice of his father, Johann F. Miescher, and his uncle, Wilhelm His, both prominent physicians and scientists, Friedrich turned to research in natural science. "It was only with the lectures on physiology," wrote Friedrich at the time, "that the entire splendor of the research on organic matters became apparent. . . ."

Miescher went to the University of Tübingen, drawn by its new faculty of natural science, and there worked under Felix Hoppe-Seyler, highly respected for his recent work on tissue chemistry. The way was open for new knowledge about cellular functions since the persistent belief in spontaneous generation was at last being dispelled. Louis Pasteur's germ theory (that living particles causing disease are transmitted through the air rather than arising out of nowhere) contributed to this, as did Rudolf Virchow's concept, published in 1858, that cells arise only from other cells. Many were fascinated with the component parts of the cell but skeptical about the possibility of separating them.

Miescher wished to study tissue chemistry and began with lymph cells, "fascinated by the thought of tracing the most generally valid conditions of cell life from the simplest and independent forms of animal cells." Inadequate supplies soon led Miescher to make pus cells rather than lymph his subject of study, a constant supply being available on the bandages of a nearby clinic. Miescher was eager to learn about the transformation of lymph cells into pus cells, what substances create tissues in the pus cells, and what the characteristics of their proteins are. Proteins, discovered about 30 years previously, derived their name from the Greek *proteios,* meaning of first importance, which everyone believed them to be at the time.

Miescher first had to separate his pus cells from the pus serum. Having no prescribed method for this or any of his other experiments, Miescher proceeded slowly with numerous tests, using varying solutions of salts. A sodium sulfate solution proved effective, and he then tested earth and alkali salts with the aim of separating the nuclear and protoplasmic substances. At this early stage he made a vitally important observation: "In the experiment with weakly alkaline fluids, precipitates were obtained from the solutions by neutralization that were not soluble in water, acetic acid, in very dilute hydrochloric acid, or in sodium chloride solution and consequently cannot belong among any of the protein substances known hitherto."

Although Miescher did not know what these precipitates were, he maintained that the substance that gave rise to pus cell material might come from the nuclei, and so determined to isolate pure nuclei. In this Miescher departed from his contemporaries, who considered the nucleus of less consequence than the remainder of the cell. His task proved extremely difficult, and initial attempts with hydrochloric acid failed. Ultimately he thought of pepsin, an enzyme that digests protein. Since the pus cells were predominantly protein, the pepsins should be effective. Miescher washed these cells with alcohol to remove some fat substances, then treated them with a solution of filtered extracts from the stomach of a pig. Pure nuclei, visible as gray sediment, separated from the yellow protoplasm. Treatment of the nuclei with an alkaline solution, followed by acid, precipitated the same substance that he had noted earlier when experimenting with the whole cell. Now Miescher was certain it came from the nucleus and he named it nuclein. He went on to find the new substance in yeast, kidney, liver, testicular, and nucleated red blood cells.

If not a protein—Miescher found it simply didn't behave like one—what was the nature of nuclein? In studying its chemical composition, Miescher detected phosphorus as well as hydrogen, carbon, oxygen, and nitrogen, the four elements that comprise most living cells. Miescher decided his nuclein was a phosphorus bank that issued this element to the cell as needed. Not only did he have no glimmering of the profound implications of his find, he had to wait two years before his work got into print, during which time a skeptical Hoppe-Seyler personally confirmed its validity.

Miescher continued his investigations back in his hometown of Basel, where he soon succeeded his uncle as professor of physiology at the university. He studied egg yolks and then spermatozoa heads from salmon, which consist primarily of nuclei and so provided a good source of nuclein. He set his students at experiments that would advance his knowledge of nuclein's chemistry. So consuming was his curiosity that at the hour he was supposed to be married, he was found in his lab.

What Miescher and other scientists at the time failed to do was to completely separate the nucleic acid of nuclein from the small amounts of protein that clung tenaciously to it. Thus their knowledge of the composition of the acids remained scant or partially erroneous. In 1889 Robert Altmann finally separated the two components and renamed the part without protein "nucleic acid." Wilhelm His maintained that his nephew had long been aware of the acid character of the substance. One wonders whether he had some inkling, too, of nucleic acid's role in the heredity of the cell. Elucidation of this would have to await scores of experiments, unexpected finds, much speculation, and intuition, as well as the more sophisticated techniques of scientists in the 20th century.

?

How did the McDonald's drive-in restaurant in San Bernardino become an international chain of 7,000 outlets?

Obviously, somebody wanted to get rich. And that somebody was not the McDonald brothers, who lived a quiet, comfortable life in southern California, content with their pretty white house and its pleasant view. A man of a different ilk proved the dynamo behind the scenes, the relentlessly driving entrepreneur who would see those now-familiar golden arches in every waking hour and in his dreams, rising majestically in every opportune empty lot. This man was Ray Kroc, middle-class midwesterner, high school dropout, former piano player and Florida real estate salesman. The son of a Western Union employee who could never get out of debt, Kroc was born to hustling and got pretty skilled at it. "It's dog-eat-dog," remarked Kroc to *Forbes* magazine in 1973, "and if anyone tries to get me I'll get them first. It's the American way of survival of the fittest." Thus in his personal credo, Kroc dismissed the age-old ideals of talent, genius, and education, endorsing instead his immense ambition.

Kroc persisted for 17 years in selling paper cups for Lily Tulip, first to pushcart vendors and Polish shopkeepers in Chicago, culminating with two soda fountain chains and a volume of 5 million cups a year. When he could go no higher, Kroc tried his hand at Multimixers, hitting the road with his little machine that could mix six milkshakes at a time. Kroc was by then in his 50s and by no means rich.

One day in 1954, Kroc drove out to San Bernardino to check out a little hamburger drive-in that, oddly enough, had seen the need to buy no less than eight of his Multimixers. On arrival,

Kroc was stunned at the volume of business (which he rapidly calculated to be about $250,000 a year). People actually waited in line (though not for long—customers could be in and out in 30 seconds) for a 15-cent hamburger in a paper bag, or a helping of French fries for a dime. Not much of a menu, but then that meant not much waste either. And no fussing with dishes and silverware. Dick and Maurice McDonald had unwittingly hit on the new American dream of dining out—fast, homogenized, clean, comfortingly predictable. It was the boon of merchandisers who could simplify, minimize, and economize. Kroc didn't intuit all this in a flash of genius. He saw only that if the McDonalds would open ten more stores he could sell 80 Multimixers.

The McDonald brothers had been approached before by entrepreneurs looking to make it big. In fact, they were already selling franchises to a few people upstate and in Arizona. But some of these did not maintain the McDonalds' high standards and this, in turn, reflected on their own establishment. Furthermore, the McDonalds were netting $75,000 and that was enough to live well on, and to do so in their hometown. That was the key. Kroc won them over because he offered a plan that would cause them the least trouble and least effort.

By the next day Kroc was on his way to Chicago with a floor plan of the McDonalds' drive-in restaurant, a recipe for their crispy French fries, and a wide-open contract allowing him to find new sites for McDonald's branches, open them—in short, run the show. Wide open with a few seemingly minor restrictions, that is, one being that all restaurants look exactly like the one in San Bernardino, with any deviation requiring written permission from the founding brothers. When the first restaurant in Illinois required a furnace and heating system, unnecessary in the original at the edge of a desert, Kroc already found himself in breach of contract. Furthermore, the financial deal would not make Kroc a millionaire overnight: he was to collect only 1.9 percent of the store owner's revenues, and of that 0.5 percent went to the McDonalds. For some years to come Kroc would have to rely on his trusty Multimixers.

The site of Kroc's first experiment was on Lee Street in Des Plaines, Illinois. At 5:30 every morning one could see an anxious middle-aged man scanning the construction, okaying every detail, and dreaming the dreams of a young man, before rushing off on the commuter train to work. Kroc had to split the ownership of the first restaurant with the construction company just to get it built. And then once it opened, horrendous problems arose with the French fries. Here in Illinois they just didn't taste like the ones out west. The recipe was tried again and again and followed with meticulous care. Even the Potato and Onion Association couldn't figure out why Kroc's potatoes were so mushy and tasteless. A scientist friend finally solved the mystery: while Kroc stored his potatoes in the basement, the McDonalds kept theirs outside in chicken-wire bins, exposed to the desert winds, which had a curing effect. Henceforth, Kroc's potatoes were cooled and cured by large electric fans in the basement.

This first restaurant, which opened in 1955, was a great success. There followed three more in California, where the market was assured. From there the rate of growth began to escalate: 37 sites by 1957 and 164 more in the succeeding two and a half years. Soon Kroc was buying out Idaho Russet Burbank's entire crop each year and dictating to the hamburger packagers just how high they could stack their patties, so his employees wouldn't have to restack them to save the bottom ones from being crushed.

In 1960 Kroc's partner Harry Sonneborn came up with a brilliant idea that would bring in hundreds of millions in revenue in years to come. He launched McDonald's in the real estate business. Now Kroc's company would lease and develop a site, and then re-lease it to the franchisee, who would have to pay rent in addition to franchising fees. Today annual rental income amounts to over $200 million, nearly 10 percent of the company's total revenue. Beginning in the late '60s, McDonald's made it a policy to buy up as many of its store properties as possible. While this initially accrued huge debts, the policy gives individual McDonald outlets the upper hand against competitors, who periodically face massive hikes in rental fees.

Kroc's commitment to his stores was total. He poured everything he had into expansion, borrowing money against future earnings. He quickly grew tired of his lethargic partners in the West, troubled by his breaches of contract, which could lead them to sue him, and anxious lest the McDonalds sell their rights to someone who did not want him around. However cash poor he still was, Kroc determined to buy them out. The asking price was $2.7 million. Sonneborn got it from New York venture capitalist John Bristol, whose clients—college endowment funds—ultimately got a $14-million return on their investment.

Then in 1963 Kroc's company went public and hundreds of millions began pouring in. The shares sold at first for $22.50, quickly rose to $50, and today go for nearly $70, after adjusting for splits.

Over the past two decades McDonald's has grown and expanded on every front: architectural variations have been made on the red-and-white-tile, box-shaped prototype; seating accommodations were added in the '60s and drive-through facilities in the '70s; Ronald McDonald began drumming up business over the radio in 1963 and by the '70s, claims McDonald's, he was familiar to 96 percent of American children; Filet-of-Fish, Chicken McNuggets, Egg McMuffins, and the wildly popular Big Mac (promoted by a $10-million "Build a Big Mac" contest) have enriched the menu. Nineteen eighty-one saw an expenditure of $322 million in advertising, making McDonald's the 14th-largest advertiser in the United States. That same year the company's university (Hamburger U) for franchisees and personnel awarded its 20,000th Bachelor of Hamburgerology. McDonald's operates over 6,700 restaurants, of which about 1,000 are in 29 foreign countries. Kroc is a billionaire and many who followed him are without a financial worry in the world. But that will not sate their hunger. Failure to grow is decidedly not the American way. And so the golden arches must line the London streets and grace the Champs-Élysées, giving the world a cheap taste of our gainful American culture.

?

How did Willie Sutton decide to rob banks?

"That's where the money was," quipped Willie.

?

And how did he do it?

At dawn on March 9, 1950, the most notorious criminal in the country, number one on the FBI's "Most Wanted" list, Willie Sutton, was getting ready for work. In his refuge on Staten Island, the Actor stood before the mirror staining his skin a dark olive, thickening his eyebrows with mascara, broadening his nostrils with two hollowed-out corks. Before going to bed he'd

IDENTIFICATION ORDER NO. 2312-A
April 27, 1951

FEDERAL BUREAU OF INVESTIGATION
UNITED STATES DEPARTMENT OF JUSTICE
WASHINGTON, D. C.

F.P.C. 12 M 1 R 000 13 / M 4 W 001
Ref: 3/4, 17/4, 19/4

WANTED

WILLIAM FRANCIS SUTTON,

with aliases: WILLIAM BOWLES, JAMES CLAYTON, RICHARD COURTNEY, LEO HOLLAND, JULIAN LORING, EDWARD LYNCH, "SLICK WILLIE", "WILLIE THE ACTOR", and others.

BANK ROBBERY

FBI No. 241,884

UNLAWFUL FLIGHT TO AVOID CONFINEMENT (ARMED ROBBERY)

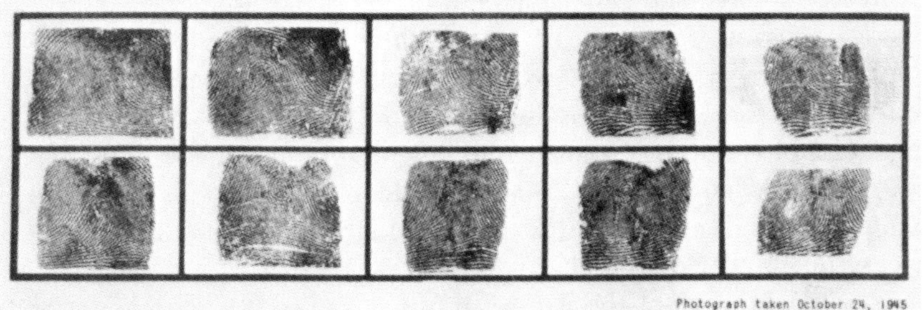

Photograph taken October 24, 1945

DESCRIPTION

Age 49, born June 30, 1901, Brooklyn, New York (not verified); Height, 5'8"; Weight, 150 pounds; Build, medium; Hair, brown, parted on left side, streaked with gray, more apparent on sides, receding at forehead; Eyes, blue, may use glasses for reading; Complexion, fair, face noticeably lined; Teeth, good; Race, white; Nationality, American; Education, eighth grade, proficient at shorthand and typewriting, speaks Spanish fluently; Occupations, clerk, driller, florist, gardener, stenographer, hospital porter; Scars and marks, faint ragged vertical scar on fold of left wrist, faint oblique scar on right elbow, end of right little finger scarred and deformed, small red flesh mole on forehead above left eyebrow, scar on back of neck, arms freckled; Characteristics and habits, smokes cigarettes in moderation, continuous user of chewing gum, dresses neatly and conservatively, reportedly wearing or carrying gloves in all seasons, soft spoken and courteous in manner, may be wearing tinted or sun glasses to disguise appearance.

CRIMINAL RECORD

Sutton's criminal record includes convictions for the crimes of burglary, grand larceny, armed robbery and escape.

CAUTION

SUTTON IS BELIEVED TO BE ARMED AND IS CONSIDERED EXTREMELY DANGEROUS.

William Sutton

A complaint was filed before a U. S. Commissioner at Brooklyn, New York on March 10, 1950, charging this subject with violating Title 18, U. S. Code, Section 2113, in that he committed the crime of bank robbery. A complaint was filed before a U. S. Commissioner at Philadelphia, Pennsylvania on April 23, 1951, charging this subject with violating Title 18, U. S. Code, Section 408e, in that he fled from the state of Pennsylvania to avoid confinement after conviction for the crime of armed robbery.

Any person having information which may assist in locating this individual is requested to immediately notify the Director of the Federal Bureau of Investigation, U. S. Department of Justice, Washington, D. C., or the Special Agent in Charge of the Division of the Federal Bureau of Investigation listed on the back hereof which is nearest his city.

IDENTIFICATION ORDER NO. 2312-A (over) Issued by: JOHN EDGAR HOOVER, DIRECTOR dyed his hair ash blond. He chose a light gray, padded suit, cut to alter his silhouette. This was a variation on the Willie Sutton trademark method of wearing a uniform that would lead the guard to open the door without the least suspicion. The additional obstacle excited Willie more than anything else had in years.

With Willie that morning was his longtime partner in crime Tommy Kling—a tough, tiny thief with a penchant for women and booze—today wearing four-inch elevator shoes for a more commanding look. They met up with the third man, John de Venuta, at a garage on Queens Boulevard where they'd stashed a stolen getaway car. Venuta had brought the Pontiac in from out of state and Willie had bought an old wreck under an assumed name, registered it, and dumped it. He kept the legitimate plates, put them on the Pontiac, then went to work on the registration—to make it match both car and license—all just in case they were stopped for a paltry traffic violation. In those days you could get a whole stack of registration forms at the motor vehicle bureau and the official rubber stamps were pretty easy to come by. But thanks to Slick Willie, the bureau subsequently changed the whole system.

The boys' target that morning was the Manufacturers Trust Company in Sunnyside, Queens. For the three preceding weeks, Willie had been there every morning studying in meticulous detail the habits of the guard, the employees, the manager. "People are, above all things, victims of habit," says Sutton in his autobiography, "especially when it comes to time." This morning he was banking on a repeat of a scenario he'd witnessed 20 times.

At 8:30 Weston, the guard, bought his newspaper at the corner of 45th Street and ambled toward the bank, his nose glued to the newsprint as usual. But this time, he'd also picked up a shadow. Slick Willie was so close behind him as he inserted the key in the door that had Weston even glanced up, he would have seen the fugitive's face reflected in the glass door. But he just pushed on in, Willie tight behind him, turning with him as he closed it, so what when Weston finally did look up, there was Sutton one step farther inside the bank than he. The stunned and trembling guard quietly acquiesced to Willie's instructions to let his partners in and to give him the grand tour—conference rooms, alarm systems, the works. Kling and Venuta set up 17 chairs facing the vault, out of sight of passersby, for the 16 employees and mailman who would soon be arriving. Poor old Weston was then latched inconspicuously to a radiator grille near the front door with a dog chain Willie had brought along for the purpose.

As each employee arrived, he or she was ushered in by Kling or Venuta and calmly informed of the situation. Willie wanted them to obey but not to panic. "Don't worry, folks," he chimed with his usual good humor. "It's only money. And it isn't your money."

Managers were a harder nut to crack, Sutton knew, because they had the authority, the status, the choice of playing the hero, with potentially disastrous results. But Slick Willie was smart. He understood what made people tick. He could play a psychological game and avoid the violence. In his long, astonishing career, the Babe Ruth of bank robbers never harmed anyone and that was why the American public loved him.

"Mr. Hoffman," Sutton addressed the manager, who arrived at 9:05, a tense four minutes late, "you're the bank manager. We know all about this place. We know what your duties are and we know what all your employees' duties are. You have the first three numbers of the combination, and Mr. Sands, your assistant manager, has the last three. You are going to open the vault for us. If you give me any trouble, I want you to know that some of these here employees of yours will be shot. I don't want you to have any false illusions about that. Now perhaps you don't care about your own safety, but the health of these here employees of yours are your responsibility. If anything happens to them, the blame will be yours, not mine."

Hoffman obliged and Sutton got into the vault. He went straight for the "teller banks," the boxes holding the tellers' cash for that day. Sands took the first step to open them with the master key, whereupon Kling brought in each teller to complete the operation, and then he slid the cash into a black silk bag. Hoffman had to open all the other compartments, including the reserve that usually has the big money. On March 9, for some reason, it was empty.

Sutton still estimated a haul of about $100,000 as he sent his partners to get the car. He ushered everyone down to the conference room, told them not to try anything, and walked suavely out of the bank. It was 9:45, just 15 minutes before the bank was to open.

The Pontiac cruised toward the Queensborough Bridge, its occupants watching police cars, sirens howling, speed in the opposite direction. They dumped the car at 72nd and York. Willie removed the plates; later he cut them up and tossed them over the side of the Staten Island ferry. The team split up and then reconvened at Venuta's place to count the take, which came to a mere $64,000. "An artistic success," remarked Willie, "but a commercial disappointment."

Although the Actor remained at large for two more years, living four blocks from police headquarters in Brooklyn, it was this job for which he was finally convicted. He hadn't really needed the cash. That very morning he'd wondered why he should be so foolish. But banks were an endless fascination for Willie Sutton. They got his blood moving. Taking control at a fevered pitch and watching his plan unfurl with precision—that made life worth living.

?

How did Shirley Temple get a contract for life insurance at the age of seven?

Only with the stipulation that no benefits would be paid if she met injury or death while drunk. Ginger ale and a maraschino cherry would have to suffice.

?

How did W. T. Grant, Inc., go broke?

At the age of 15, William T. Grant left school to help support his mother by running messages in a New England store. The ambitious youth soon purchased an interest in the business and proceeded to buy out his partners. The ensuing years saw this small-town store swell to a massive, nationwide variety-store chain, third in size after F. W. Woolworth Company and S. S. Kresge Company. In 1975 the Grant empire comprised 1,100 stores and 75,000 employees. But that same year brought unprecedented disaster: the largest retailing bankruptcy in United States history to date. According to *Fortune* magazine, Grant's debt totaled $800 million. Profits had declined drastically over the previous five years, culminating in massive operating losses in 1974 and the first seven months of 1975. As reported by *The New York Times,* net income slipped from $39.6 million in 1970 to a loss of $177.3 million in 1974. Grant filed for bankruptcy in October 1975, and within three months closed 712 stores and dismissed 50,000 employees. In March 1976 the Grant logo in 20-foot-high electrical letters crowning 1515 Broadway was dismantled. The company abandoned its nine and a half floors of store space for humbler, temporary lodgings at 360 West 31st Street. A Los Angeles firm, Sam Nassi Company, oversaw the liquidation of Grant stores nationwide, and crowds stampeded as the stores slashed prices by 60 percent and more.

How, experts queried, could a multimillion-dollar chain of Grant's stature run amok so completely and irrevocably? Initially, the world supposed that the company had simply overreached itself and failed to maintain an ample and loyal clientele. But investigations in years subsequent to the company's downfall exposed a maze of internal problems. As reported by *The New York Times* on February 4, 1977, former executives and employees testified that:

—Buyers often had to ask vendors what Grant's inventories were.

—Vendors were deliberately asked to overbill the company and to keep the unrecorded money until it was requested, often being paid back in checks made out to the chain's buyers.
—The merchandise vice-president hadn't any buying budget.
—Each store manager had his own system of consumer-credit checking.

Overbilling was perhaps the most grievous problem, costing the company millions. Under this system—a common and fully legal one among large retail chains—Grant would contract in its New York office to buy goods at *x* dollars, the going rate for those goods. It would then have the vendors bill Grant for *x* plus *y* dollars. The *y* amount would be held by the vendor (in effect, an interest-free loan) until collected by Grant, supposedly in the fourth quarter. The overbilling account, known as 1517, served two purposes, according to Richard H. Rosenblum, former Grant assistant general counsel. It provided funds for racks at Christmastime and other administrative items, and made sales and profits in the fourth quarter appear greater than they actually were. A Federal Trade Commission source observed, too, that "an obvious purpose of the practice could be to gain a lever with suppliers by letting them use the overbilling money."

Obviously, such a system demanded rigid controls and meticulous organization, both of which were lacking at Grant. Apparently, the retailer had no orderly records of how much was in the 1517 account and relied on individual buyers to keep track—which they did only occasionally. Furthermore, if a buyer left or a vendor stopped dealing with Grant, the company had no way of knowing the amounts to be collected, nor, since there was nothing in writing, could they force the vendor to pay up. When Robert Anderson, a former top executive at Sears, Roebuck, became Grant's president in the spring of 1975, he tried to stop the overbilling system and claims that he did so. But in the process, the company had to write letters to their various suppliers and ask *them* how much was owed.

A massive credit drive was another ill-advised idea that sapped the company's resources. Pressure on managers was made all too real by a "Steak and Beans" program of "negative incentives," in which those who failed to meet their credit quotas suffered humiliating hazing not unlike that at college fraternities. Some had custard pies smashed in their faces and their ties cut in half; others had to run around their stores backward or push peanuts with their noses; one failed manager allegedly was forced to promenade about a hotel lobby in nothing but a diaper. Naturally, most district managers preferred to avoid this sort of experience and so gave credit to "anybody who breathed," according to one former executive. Since managers themselves had final approval on new accounts, this approval was forthcoming with only the vaguest attempt at discovering whether the new client was responsible; the rate of rejection was about 20 percent as compared with 80 percent in 1974 when controls were tightened. The credit drive produced results: between 1967 and 1974 credit sales soared from 15 percent of the total volume to 30 percent. But by the early '70s, it was common knowledge that the best place to furnish your home on credit was at Grant, because if you didn't wish to pay, chances were you wouldn't have to. Uncollectibles rose from 2.1 percent in fiscal 1970 to 3.2 percent in fiscal 1972. The company's computers were not yet sophisticated enough to accommodate large credit programs and coordinate accounts between stores. So while some customers merrily charged huge sums at several stores and neglected to pay, hundreds of others found the intense pressure to open an account so unpleasant that they swore never to set foot in a Grant branch again. All told, the credit system was a disaster. In June 1976, eight months after the company's demise, the trustee reportedly sold $276 million in uncollected customer receivables for a paltry $44 million.

?

How did they discover Neptune?

It was not the searching eye of an astronomer but the enormously complex calculations of mathematicians that made Neptune known to the world. If skeptical at first, people were dazzled when the new planet was finally seen and the computations thus confirmed. This was the first planet to be discovered mathematically, and the method seemed miraculous.

Two brilliant young men, independent of and unknown to each other, addressed the problem in a similar fashion, with similar results. What sparked their undertakings was an attempt to explain mysterious perturbations in the orbit of Uranus. Ever since this planet's discovery by William Herschel in 1781, astronomers had been puzzled by its irregular orbit—it did not act as it should if Newton's principle of universal gravitation was to be believed. Some suggested the presence of a mysterious cloud or gas that was detaining the planet. Others believed a comet had collided with Uranus and thrown it off kilter. Still others decided that Kepler and Newton must simply be wrong. But many belived there might be a distant planet whose gravitational pull was responsible for Uranus's course. No one, however, believed the position of the planet and its size could be computed, and anyway the task was for years prohibitively complex.

Then in 1843 a phenomenally talented Cambridge mathematician, John Couch Adams, having just received his degree, began work on the problem. By September 1845, his intensive work paid off—he had figured the orbit and size of the new planet. Now he needed confirmation from the astronomers. Adams sent his findings first to the astronomer royal, George Airy, who cast an arrogant eye on the work of the 26-year-old, then to James Challis, director of the university observatory, who likewise showed little interest. The two simply could not rely on theoretical deductions, and the mathematics was undoubtedly too much for them. Airy, furthermore, thought Adams had made a good guess at the orbit, then considered whether the perturbations of Uranus, as observed, would follow. But Adams, of course, had approached the problem the other way around—from the perturbations to the orbit.

In France, a respected mathematician, Urbain J. J. Leverrier, 34, also tackled the complex problem and came up with a nearly identical solution in June 1846. He published his results, giving the same position of the planet, to within 1°, as that reported by Adams. Challis and Airy noted this with some astonishment, but even then they were slow to act. Airy asked Challis to begin the search with the refractor at the Cambridge observatory, one of the largest telescopes existing at the time, with its 25-inch-aperture object glass. Challis's tragic error was that he would not focus solely on that position designated by Adams, but insisted on searching an area of 30° wide and 10° deep. After two months he had measured some 3,000 star positions, but neglected to chart them and so failed to note the movement of a planet in their midst.

On the Continent, Leverrier was eager for recognition. Finding the French astronomers as hopelessly stodgy as the English, he sent his computations to Johann Galle, director of the Berlin Observatory. On the very night he received Leverrier's material, September 23, 1846, Galle turned his refractor (with a nine-inch aperture) to the specified position in the sky—only to find a clutter of stars. The search might have quietly ended there, but Heinrich d'Arrest, an assistant, recalled a chart of that portion of the heavens that had been prepared by the observatory two years previously but never published. When they compared the stars before them with those on the chart, they found something new. With much excitement they watched all night, yet couldn't observe enough movement of the body to be sure it was the undiscovered planet. But the next night Galle saw all that he needed to: the dim irregular disk had moved the predicted amount and in the proper direction, and lay within a degree of the position that Leverrier had specified. This, then, was the eighth planet, Neptune, orbiting the sun at a mean distance of 2,796,700,000 miles.

As might be expected, a heated debate raged across the English Channel about who should take the credit. After some initial injustices to the more reserved John Adams, the world ultimately decided that he and Leverrier might share the honor equally.

?

How did they draw the Mason-Dixon line a century before the Civil War?

Since the 19th century, the Mason-Dixon line has represented the demarcation between North and South, the front where opinions clashed on the incendiary issue of slavery. In Civil War days, the line ran along the Maryland-Pennsylvania border and westward along the Ohio River to the Mississippi. But Charles Mason and Jeremiah Dixon actually had no intention of charting the West. Their sights were closer to home. Between 1763 and 1767, they drew a line along the Pennsylvania-Maryland border, and only to settle a long-standing family feud between the Penns to the north and the Calverts to the south.

?

How did the whooping crane learn to reproduce in captivity?

The fight to save the endangered whooping crane began as early as the '30s. Since then the struggle has acquired a certain mystique, exemplifying the recent awareness that man cannot live at the expense of other creatures. The United States government and the National Audubon Society initiated studies and programs to aid the majestic cranes, and National Audobon Society Research Director Robert Porter Allen in 1952 published a monograph *The Whooping Crane*, which became a classic on the bird, whose habits were until then little known.

The tallest flying bird in North America, the whooping crane stands up to five feet tall, with a wing span of nearly eight feet. The bird is white, with spindly black legs, silky black wing tips, a sinuous neck, and a vibrant buglelike voice. The single wild flock, numbering 73 in 1982, migrates thousands of miles between the Wood Buffalo National Park in Canada's Northwest Territories and the Aransas National Wildlife Refuge on the Texas Gulf coast.

One of the most ingenious attempts to rehabilitate the cranes was that of Dr. George Archibald, who in a sense became the "father" of a baby crane. MAN AND BIRD DANCE TOGETHER TO PRESERVE SPECIES reported *The New York Times* in March 1980. Archibald had, in fact, mastered the ritual mating dance of the whooping crane and performed the lively *pas de deux* with Tex, a 13-year-old crane living at Archibald's International Crane Foundation in Baraboo, Wisconsin. Tex (who was christened in San Antonio, Texas, before her sex was determined) might more suitably have danced with the dashing male in a nearby pen, but she had been raised by humans and had developed a filial attachment to people rather than to her own species. So for three spring mating seasons running, Archibald and Tex had danced and flapped and called to each other. Starting about eight feet from the regal crane, Archibald raised his arms, jumped straight up, then crouched with arms still outspread. He continued doing deep knee bends, occasionally picking up grass and sticks and throwing them in the air—all of which stimulated Tex to join him. She succumbed, shimmying and huffing, lifting her neck up and down, turning and flapping her wings, and sometimes picking up grass and sticks. Then when she stretched her wings and lifted her tail in the climax of the ritual, three attendants rushed into the pen and artificially inseminated the receptive crane. They inserted a syringe in the bird's cloaca and injected sperm obtained from a male whooping crane that morning.

Tex produced one egg each spring but these three never added a healthy chick to the fragile whooper population: one was infertile, another contained a deformed chick that soon died, and the shell of the third was too thin. In 1980 Archibald was unable to dance, and Tex, it appeared, would have no other. She refused to dance with Japanese ornithologist Yoshimitsu Shigeta, who even wore a special white jumpsuit to enhance his image. Whooping cranes are, in fact, monogamous throughout their long lives of 30 to 60 years, and Tex was hooked on Archibald. In 1982 the amorous pair were reunited. Archibald spent 12 to 16 hours a

Courtship at dusk. Dr. George Archibald inspires Tex to join him in the ritual mating dance of the whooping crane.

day in her pen for six consecutive weeks, and Tex at last laid a fine, strong egg. On June 1, a healthy chick hatched and was aptly named Gee Whiz.

Two months later a violent tragedy ended the romance and plans for future progeny. A raccoon climbed the fence around Tex's pen, gnawed through the flight netting, and killed her, leaving only Gee Whiz to live up to his mother's fame and carry on her line.

?

How did Louis XIV dine?

From the moment the Sun King arose from his sumptuous gold bed, aligned with the rising sun at the centerpiece of his beloved Versailles, the château was alive with activity. The life of every courtier, minister, lovely lady, doctor, and cook was finely tuned to the rituals of the King—his dressing, shaving, dining, meetings, and evening *comédie,* dancing or *appartement* when the halls were flooded with light and the courtiers played billiards, gambled, and ate sweets. Living under a pretense of usefulness to Louis XIV and the government of France, the courtiers were largely a source of amusement to the King, whose favors and beneficence were vital to their well-being. "*L'état c'est moi,*" said Louis with solemn conviction, and Versailles was a convincing illustration.

Louis XIV's taste for massive architecture, yawning vistas, and spacious gardens was reflected in the realm of gastronomy, for his appetite was equally huge. "He had never known hunger," reported the Duc de Saint-Simon, who lived in Versailles for 20 years, but "I heard him say several times that his appetite was whetted with the first spoonfuls of soup, and he ate so prodigiously, so substantially, and so equably morning and evening that one never tired of watching him." The King arose at about 8 or 9 A.M. and, after attending mass and meeting with some ministers, sat down for the first feast. At 2 P.M. a procession of dishes arrived and was served in his bedroom, where he sat gazing out on the gardens and the public, who wandered freely through the grounds and many of the chambers. "I have often seen him," said Mme. Maintenon, his second wife, "eat four plates of soup, a whole pheasant, a partridge, a large plate of salad, mutton au jus or à l'ail, two big slices of ham, a whole plateful of pastry, and, besides these, fruit and hard-boiled eggs." She further commented that if she were to consume half this amount she would be dead in a week. The King devoured oysters by the dozen, supplied by his own private park, overseen by a man named Hyacinth Ox. He ate all sorts of fish, despite the fact it was transported over great distances and was not always in a state we would relish today. For a fresh supply of citrus, he maintained a grove of over 1,000 rare orange trees, planted in silver tubs.

"Everything [the King] ate," recalled Saint-Simon, "was at least twice as highly spiced and pungent as is customary. Fagon [the doctor] was against sweets and spices and when he saw the King eat them he made very amusing faces. Yet he dared not say anything except from time to time to Livry and Benoist [the cooks] who replied it was their job to feed the King and his job to purge him." Indeed, the King suffered from dyspepsia and later from gout. Others at court succumbed to gastric ulcers and frequent dyspepsia; they, unlike the King, also gorged on pastries, preserved fruits, cider, and lemonade *between* the lavish meals.

The King supped at 10 P.M., occasionally in public. On such occasions hordes drove from Paris—past peasants who during the famine of 1705–1708 lay dying by the road with only weeds and tree bark to eat—to file before the King and behold the spectacle of his feast. The courtiers competed for the honor of serving him, which entitled them to eat the leftovers but demanded that they sample each dish before it entered the King's mouth. For the second time in a day, there were boundless quantities of soup, meats, fish, salad, ripe fruit, and the inevitable hard-boiled eggs. Some 64 dishes—8 courses with 8 choices each—were prepared and passed before the King, from which he selected about 20. To the long, often fruitless hours of preparation must be added time spent bringing the meals to Versailles. In those days the château had no kitchens (or bathrooms): the food was prepared *outside* the château in the village and carried to the King, by which point it was all cold and either served that way or reheated on special burners. One can well imagine the resentment of the servants in the dead of winter, or of the hungry peasants who smelled the gastronomic delights as they passed by, and marvel that such an order was possible and would persist for decades to come.

Day after day Mme. Maintenon sat with the King, dressed in his habitual brown robes and undercoat studded with precious gems, as he waded through his gargantuan meals, hardly uttering a word. "It seemed," she later remarked, "as if he had limited himself to a fixed number of words in his lifetime and was afraid of exceeding the limit." Whatever finery he wore, the King always ate with his fingers, and Saint-Simon marveled at his ability to eat chicken stew without spilling a drop. Individual forks had come into use only a century before in Italy and most European countries had not yet adopted them. Individual knives and spoons were now often supplied, whereas before the 16th century a communal spoon sufficed and one might bring one's own knife. Louis resisted the fork, however, forbidding the Duke of Burgundy and his brothers from using it at his court, although they had been taught to do so.

When Louis XIV died in 1715 and the customary autopsy was performed, all were stunned at the size of his stomach and intestines, which was twice that of a normal man's. It was discovered, too, that his enormous gut contained an enormous tapeworm. This, however, was not responsible for his voracious appetite. If anything, tapeworms can cause a *loss* of appetite. Who knows what might have transpired had the King ever been hungry.

?

How did they decide the length of a mile?

The statutory mile (as opposed to the slightly longer nautical mile) has an unlikely and seemingly arbitrary length of 5,280 feet. The measurement derives from ancient Rome. As walking was then the primary mode of transportation, the Romans measured a distance in paces—2 steps to the pace, a pace spanning about 5 feet—thus *milia passum,* 1,000 paces, measured 5,000 feet. This Roman mile seems logical and might have prevailed had not the British Parliament, which established all kinds of measurements over the years, decided in the 16th century to tack on some more footage. Parliament announced the mile to be 1,760 yards or 5,280 feet, so that it could be divided evenly into furlongs, the most common land measurement of the time. A furlong was the length of a plowed furrow in an ordinary field—660 feet—and thus there were 8 furlongs to the mile. The purpose of the change was to facilitate surveying, and Elizabeth I proclaimed it the law in 1575.

?

How did they build the Great Pyramid at Giza?

No one throughout the ages has enjoyed so sublime a tomb as that of King Khufu, builder of the magnificent Great Pyramid of Giza. The son and successor of King Snefru and Queen Hetepheres, Khufu reigned for 23 years, a period marked by economic stability and artistic progress. The able and energetic King erected several monuments in Egypt and in later years his name, found engraved on scarabs, apparently assumed a potent, protective charm. Little else is known about this second king of the IVth Dynasty (2613–2498 B.C.); indeed, there is no evidence to support the claims of classical authors that he was a tyrant, abhorred by all compelled to labor on his behalf. It is more likely, according to Ahmed Fakhry, author of *The Pyramids,* that these subjects were more than willing to work for the eternal glory of their ruler, who in their eyes was a god.

The Great Pyramid culminated in both size and quality a long tradition of pyramid building. The awesome edifice, perhaps chief of the Seven Wonders of the World, gives pause even to modern-day architects and engineers who have 5,000 years of experience on the ancient Egyptians and the aid of advanced technology.

Without wheels, pulleys, compasses, or tools of any metal other than copper, the Egyptians raised a pyramid almost perfectly aligned with the cardinal points, covering an area of over 13 acres, and piercing the sky at a height of 481.4 feet. An estimated 2.3 million blocks of stone make up the Great Pyramid. Weighing an average of 2½ tons each, some as much as 15 tons, the limestone and granite monoliths were cut, hauled, and positioned with enormous precision. St. Paul's Cathedral and the Houses of Parliament could stand within the perimeter of the base, with ample room to spare. And if all the stones in the pyramid were sawed into blocks one foot square and laid end to end, the line would stretch two thirds of the way around the earth at the equator.

Extant Egyptian texts say all too little about the construction of the pyramids, and authorities continue to puzzle and disagree over the methods employed. What is known has been surmised from the pyramids themselves and some knowledge of the type of implements available to this extraordinary society.

The first task for King Khufu and his engineers was the selection of an appropriate site. Certain strictures always applied. The tomb had to lie on the western bank of the Nile, for the realm of the dead lay in the west, where the sun went down. It required a strong supportive base of solid rock and sufficient room for the surrounding mortuary temple, causeway, valley temple, and canal connecting with the Nile. (East of the Great Pyramid stand three small ones, erected for the King's wives or possibly for a wife and two daughters. Herodotus recounts a spurious story that the central pyramid was built by one of Khufu's daughters, whom the King prostituted in order to gain funds for his own tomb. The young lady allegedly asked each lover to contribute a stone and so looked after her own immortality.)

King Khufu chose a plateau, rising 200 feet above the river, near the ancient capital of Memphis; it is part of the Memphis cemetery extending from Abou-Rawash in the north to Meidum in the south. His workers removed the thick surface layer of sand and gravel, so that the pyramid might stand on the firm rock substratum. Next they leveled and smoothed the rock, a technique learned gradually over the years from preparing lands for irrigation. The workers built four banks of Nile mud around the area to be leveled and filled the enclosure with water. They then cut a maze of trenches, each lying at precisely the same depth beneath the water's surface. Later, after the water was run off, the areas between the trenches could be leveled to them, The system produced superb results: the perimeter of the pyramid's base deviates from true level by just over half an inch. The area within did not have to be absolutely level with this perimeter, and in fact a mound of rock was left in the center and used later.

The principal architect, HemIwnw, Khufu's cousin, then had to align the four sides of the pyramid with the four cardinal points. Here King Khufu himself might have been seen ceremoniously drawing the first line and assuming the power of this wisdom before his people, but it was undoubtedly the architects and astronomers who did the tricky calculations—all without the aid of the magnetic compass. Their errors amount only to fractions of a single degree. (The north side is 2′28″ south of west, the south side 1′57″ south of west, the east side 5′30″ west of north, and the west side 2′30″ west of north.) Could the architects have observed the sunrise and sunset on the two annual equinoctial days, or taken an accurate reading of the north by observing the Pole Star? I.E.S. Edwards in *The Pyramids of Egypt* maintains that in both cases the resultant error would have been far greater than that found at the Great Pyramid.

For a description of the foundation ceremonies and implements used, historians are forced to look to texts written many years after the pyramids were built, and speculate that the methods were fundamentally the same. Two devices were employed: the *merkhet* and the *bay*, both mentioned first in the texts of the XVIIIth Dynasty. The *merkhet*, which means instrument of knowing, consisted of a narrow horizontal bar with a small block rising above the level of the bar at one end, from which a plumb line was suspended. This was used with a *bay*, or palm rib with a V-shaped slot in the wider end. A likely approach was to observe a star in the northern heavens and bisect the angle formed by its rising position, the point of observation, and its setting position. For an accurate reading, the Egyptians would need a true horizon; since their natural one was distorted by the line of the land, the architects had to build on the leveled plateau a circular wall whose upper rim would serve as a horizon. One person, positioned at the center of the circle, would sight through a perpendicular rod or slot of the *bay*. When he saw the rising star he would instruct another person to mark with the *merkhet* the exact spot on the wall, i.e., the direct line between the observer and the star. Some hours later another observation would be made, this time toward the west, and the place indicated by another *merkhet*. The two plumb lines, suspended from the wall, would indicate two points perpendicularly below, at the base of the wall. By bisecting the angle formed by these two points and the observer, the Egyptians could find true north. The reading might be checked by observing other stars in a similar way. Once one axis was established, the other could easily be found by marking off a set square.

The four sides of the Great Pyramid were originally just over 755 feet each at the base (now slightly shorter because of lost casing stones), with the greatest difference between them only 7.9 inches. This error probably resulted from the cords used in measuring the base, which were of palm fiber or flax fiber and naturally stretched when used. Because of the mound in the middle, the Egyptians were unable to measure the diagonals to double-check their work; still, they were remarkably accurate.

While all this was going on, gangs of workers were already toiling away to obtain the staggering volumes of stone that the King desired. From inscriptions painted on some of the rocks we know that each of these gangs had a name, e.g., "The Craftsmen-gang. How powerful is The White Crown of Khnum Khufu!" Marking the stones may also have provided a way for overseers to check the workers' productivity. The stones are predominantly limestone, taken from the plateau itself and nearby quarries with nothing but copper chisels and saws, flint tools, quartz and diorite pounders, and wooden crowbars. The gangs cut away three sides of each stone with a chisel struck by a wooden mallet. They then inserted wedges into holes cut at the bottom and freed the rock from its base. Sometimes these wedges were of wood, which expanded when wet and thus caused the rock to split. This process was made yet more difficult when tunneling was required, as at Tura on the eastern coast of the Nile, where finer-grained limestone was quarried for the outer casing. Here confined space severely restricted mobility and the number of workers.

Historians continue to debate how the Egyptians of the Old Kingdom were able to quarry granite, a much harder stone than limestone. One authority insists that they could not do it at all and simply used what was found on the ground. Some maintain that wedge slots could have been made by rubbing an abrasive powder with a stone or metal tool; others say the workers pounded around the desired rock with balls of hard dolerite, a stone widely distributed over the eastern desert near the Red Sea.

The Egyptians then had to transport their stones, something they proved to be remarkably good at over the years—later moving the 1,000-ton colossus of Ramses II to Rameseum at West Thebes. Megaliths from Aswan and Tura were floated on barges during flood season. Overland transport was effected without wheels, in much the same way as at Stonehenge (see "How did they get the stones in place at Stonehenge?"). Workers levered the stones onto sledges or dragged them on via short stone ramps. They then raised both stone and sledge, bound securely together with ropes, enough to slide wooden rollers underneath. Straining on ropes attached to the sledge, the men dragged the sledge along a way lined with logs, while others poured water or milk over the timber to prevent fires resulting from the friction. Offerings may well have been placed on the stones and incense burned as the gangs sweated over their slow and painful work. Herodotus reports that 100,000 men were employed for periods of three months just for the task of transporting the stones, but these figures have not been confirmed. He claims, furthermore, that the causeway required 10 and the pyramid itself 20 years to build, somehow overlooking the fact that Khufu reigned for only 23.

More certain is the number of full-time workers engaged in building the pyramid. Sir Flinders Petrie, who did the first exhaustive survey of the pyramid in 1880–1882, discovered barracks lying west of the Pyramid of Chephren and presumed that they once housed these skilled workers and could have accommodated about 4,000. In 1978 and 1980 Zahi Hawass, chief inspector of the Giza pyramids, excavated a site northeast of the Great Sphinx and found under the modern village of Nazelet-el-Saman a village called Busiris dating from the Greek and Roman period. He also found Old Kingdom deposits suggesting that at this site there was once a workers' village from the time of Khufu and even a palace for the King. Thus Khufu not only oversaw construction of his tomb but ensured ample food, clothes, and shelter for thousands in his employ. His organizational abilities were yet another awe-inspiring factor in the building of the Great Pyramid.

Once the stones had been dragged to the plateau, a skilled group of workers smoothed, shaped, and assembled them. After the first tier, their work was considerably harder. How the stones were raised without pulleys has puzzled historians, architects, and engineers for centuries, and the definitive word has not yet

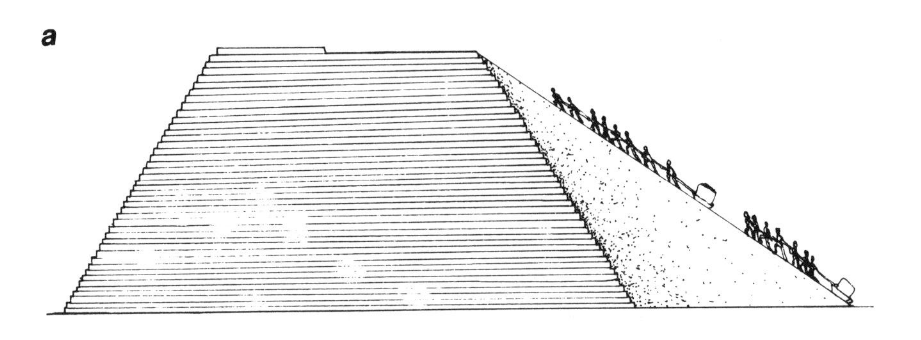

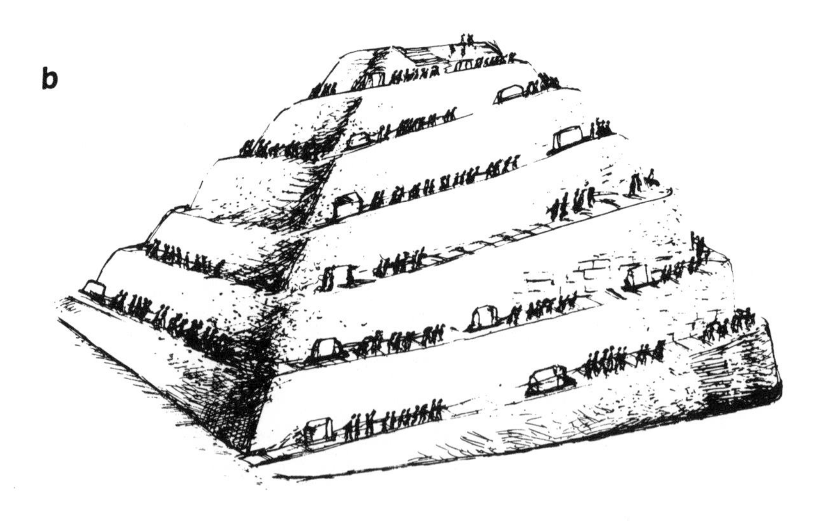

To get the materials up the pyramid as it rose, the Egyptians may have constructed an inclined ramp against one face (a) or slanting side ramps (b), one starting at each corner.

been written. Most now agree, however, that the only feasible method was the use of ramps, and tangible evidence appears at the unfinished pyramid of King Sekhem-khet at Saqqara, where the rubble of such ramps was found. In *Ancient Egyptian Masonry,* architect Somers Clarke and engineer R. Englebach outline a system employing both long, wide supply ramps with a fairly easy gradient and shorter, steeper ramps for laborers to transport lighter materials. Edwards suggests that a single supply ramp was constructed over an entire side of the pyramid and raised successively as the monument went up. Like that of the pyramid itself, the ramp's angle would have been 52°. On the other three sides, meanwhile, foothold embankments would have been formed, sufficiently wide for men to navigate with building materials.

Other Egyptologists argue that a single long ramp was unfeasible because of the difficulty of raising and extending it at each level. In 1950 Dows Dunham, curator of Egyptian and Ancient Near Eastern Art at the Museum of Fine Arts in Boston, oversaw construction of a model of the third pyramid at Giza on a scale of 1:120. This model incorporated four ramps, one starting at each corner. Dunham maintained that the three were used for delivery of stones and the fourth reserved for men descending with empty sledges. Each ramp would have been about ten feet wide, sufficient for the sledges, a double row of men to haul them, and wet timber laid like railway ties to reduce the friction.

Imagine the pyramid in the midst of construction at about half its present height, the top a flat, square surface. In the broiling sun, thousands of workers mill around its base. On the ramps, they are hauling the core stones from local quarries; these are left rough on all sides except the bottom. Placed side by side, the gaps caused by their uneven surfaces are left open, but careful measurements are made to keep the sides of the square of equal length. Sometimes internal casing of Tura limestone is added to make up the difference. Finally, only a narrow margin remains at the outer edges of the square platform. Skillful workers now position the limestone "packing blocks" with close-fitting joints. The task becomes still more delicate, for it is time to add the outer casing of Tura limestone. Expert masons shape the precise angles of these stones on the ground, perhaps fitting them to the packing

Closely fitting casing stones of white limestone, seen here on the lowest level, once covered the Great Pyramid of Giza. Today, the large underlying blocks, cut with flint and copper tools, are exposed.

blocks before these are hauled away. The masons leave bosses on the outer face of every casing block so that levers may be used. They spread a thin layer of mortar on the bottom and inside surfaces, which facilitates sliding the stone into position. Once the stone is hauled up the ramp and laid on the casing block below it, men position it precisely by pulling on ropes attached to its outer corner, then by levering it from the front against the packing block.

Petrie took a close look at the few casing blocks that remain at the Great Pyramid, and his report of the accuracy of measurement is absolutely incredible: "The mean thickness of the joints of the north-eastern casing-stones is 0.02 inches, and therefore the mean variation of the cutting of the stone from the straight line and from a true square is but 0.01 on a length of 75 inches up the face, an amount of accuracy equal to the most modern opticians' straight-edges of such a length. Though the stones were brought as close as 1/50 inch, or, in fact, in contact, the mean opening of the joint was but 1/100 inch."

Once the casing stones are on all four sides, a survey is made. Ramp and embankments are meanwhile raised and the entire process repeated. After many, many years of labor, the workers one day haul up on a sledge the pinnacle of the pyramid—the capstone. They support it on levers and place it on battens. The square surface at the top has been covered with mortar and a small depression cut to hold a protruding disc on the bottom of the capstone—a mortise and tenon system to keep the stone in place. Finally, they remove the battens and secure the lofty pinnacle. Even now the work is far from over. They must dress the stone as they descend, and gradually dismantle the ramp. Then they must clear the grounds and prepare to build the mortuary temple, the causeway corridor, and the valley temple.

As a footnote to this Herculean effort, it must be recalled that the Great Pyramid contains a maze of interior passages, the Queen's Chamber, the King's Chamber, five "relieving chambers" (designed to take much of the weight of the upper pyramid from the roof of the burial chamber), and the majestic Grand Gallery, a corbel-vaulted passage 153 feet long. The details of construction are too many to describe here, and much remains unknown. It is probable, however, that this work was done independently of that on the exterior. Numerous small ramps may have been erected at various stages of construction, so that the interior chambers could be built above the core construction and completed before the surrounding pyramid precluded access. (Later, entrance through the passage on the north face, via a supply ramp, would be possible.) To expedite construction, many of the stones were smoothed and sized on the ground outside. The roof slabs of the King's Chamber were numbered as well, so that workers inside could position them easily and quickly.

When King Khufu died, his divine body was washed and purified at the valley temple and then mummified, a ritual that may have taken several months. (An inscription at Queen Mersyankh's tomb in the eastern cemetery at Giza records an interval of 272 days between death and burial.) There followed a ceremony called The Opening of the Mouth, a magic rite performed on the burial day that would allow the King to speak once more and enjoy the offerings in the life to come. The mummy was carried to its luminous white tomb and laid within, the entrance closed forever and concealed behind casing stones. Now cults of royal priests, dressed in white robes, began rites and services that did not end until Ptolemaic times. They heaped the altars with flowers, presented offerings each day, and filled the temple halls with their sacred hymns and prayers.

?

How did Baudelaire's *Les Fleurs du Mal* get reviewed when it was published in 1857?

His mother had reservations.

Théophile Gautier warned the poet of its "scabrous nature."

M. Bourdin, writing for *Le Figaro* (a newspaper largely reflecting government views), was scandalized:

> There are times when one doubts Monsieur Baudelaire's sanity; there are those when there is no longer any doubt; most of the time we have monotonous and calculated reiterations of the same words and notions—the odious is cheek by jowl with the ignoble—the repulsive joins the disgusting. You have never seen so many bosoms being bitten, chewed even, in so few pages; never has there been such a procession of demons, fetuses, devils, animalia, cats and vermin.
>
> The book is a hospital open to all forms of mental derangements and emotional putrefaction. . . .
>
> If one may understand a poet at the age of twenty allowing his imagination to be carried away by such subjects, nothing can justify a man of over thirty making public such monstrosities in a book.

French courts banned six of the poems and fined the poet and printers on the grounds that the poems were "necessarily conducive to the arousal of the senses by virtue of a coarse realism offensive to modesty."

While Baudelaire considered scandal "the foundation of [his] future," he was antagonized by the court's ignorant and unperceptive accusation of "realism." Spirituality was the very essence of his artistic vision, and this alone, he maintained, transformed the squalor and evil of the world into beauty.

?

How did they control the population before the Pill?

There may be just one basic way to conceive a child, but people over the ages have concocted an array of ways not to. The human race has seen fit to control its numbers for diverse reasons—be it shortages of food in Neolithic times or eugenics in ancient Sparta—and the means to this end have been ingenious, baffling, and bizarre.

In the days of our Paleolithic ancestors it is likely human lives were numbered largely by external factors, over which they had no control. For one, the mortality rate was extremely high. The population of the world 1 million years ago has been estimated at half a million, and only 3 million in 10,000 B.C.—990,000 years later—suggesting long periods of zero population growth. Disease went unchecked and malnutrition depressed fertility, as well as resulting in infant and child deaths. The average life span, furthermore, was a fraction of what it is today: in Neanderthal times, 70,000 years ago, only about two strong and cunning persons out of ten reached 30, the age at which many women today begin thinking about a family. Not only were childbearing years reduced, but women in general were few, outnumbered by men three to two. This suggests the practice of infanticide, for surely it would have been the female, the future bearer of unwanted children, who would be killed, rather than the male, a more useful hunter and warrior.

In Paleolithic times, no steps would have been taken by the male to prevent procreation (a situation not unlike that today) because no one realized the male's role in the creation of new life. But it is possible that women tried to assume a degree of control themselves. As Reay Tannahill points out in *Sex in History*, women undoubtedly knew a great deal about the medicinal properties of plants and may have ingested herbal and vegetative brews to prevent pregnancy, as various peoples do today. In the central forests of Paraguay, for instance, some women drink an infusion derived from a plant, *Stevia rebaudiana*, to avoid pregnancy. The Navaho have used a tea of ragleaf Bahia, the Hopi a powder from the root of jack-in-the-pulpit. There is some scientific evidence from experiments with animals that such medicines achieve the desired effect.

If we jump tens of thousands of years to the sophisticated civilization of ancient Egypt, with all its advanced technologies, sciences, and arts, we find that these people were also savvy about sex. They understood that the seminal fluid must be blocked from entering the womb if one did not want to get pregnant. For this purpose they devised some singular recipes, which date from dynastic times. The Kahun Papyrus advises women to mix crocodile dung with a paste called *auyt* (as yet unidentified) and to insert this in the vagina, while another prescription calls for a compound of honey and natron (sodium carbonate). The later Papyrus Ebers recommends soaking a lint pad in a mixture of acacia tips and honey and positioning this (no mention of how) at the cervical opening. In this the Egyptian chemists outdid themselves, for we now know that sperm favor alkaline rather than acid conditions and that acacia tips produce lactic acid. This is the active ingredient used in contraceptive jellies today.

Abortion, infanticide, the ingestion of drugs, homosexuality, zoophilia, and coitus interruptus were among the other methods early civilizations used to keep the population in check. This last presented a problem for the Hebrews, however, since the Torah instructs men to be fruitful and multiply. Thus the onus fell on the women, and under certain situations a contraceptive sponge was compulsory: for girls 11–12, nursing mothers, and pregnant women (to avoid a disruptive second conception, which the Hebrews believed could occur). Some wives jumped up and down to get rid of semen. Others drank "the cup of roots"—Alexandrian gum, liquid of alum, and garden crocus—mixed with two cups of beer.

Women fared somewhat worse in ancient Greece, for this civilization, however learned and illustrious, did not look kindly upon them. Their political and legal rights were scarcely better than a slave's, and in countless instances even the right to live was denied. For many years the Persian and Peloponnesian wars drained the country of its young men, and the balance of the sexes would have tipped heavily toward the female if something had not been done. Here the solution was infanticide. In Sparta, where the goal was a stronger and improved race, feeble infants of both sexes were left to die in the Taygetus Mountains.

For the most part, though, overpopulation was not a problem in Greece. Men's disdain for women and preference for each other kept population growth down, and the state in Crete encouraged pederasty for precisely this reason. If families did start getting too large, Aristotle advised, Greek women should apply olive oil blended with cedar oil, lead ointment, or frankincense to "that part of the womb in which the seed falls."

Olive oil was a favorite with the women of Rome, too. Small wonder when one considers some of the other proffered means. Pliny tried to nip the problem in the bud simply by curbing a woman's sexual desire, enjoining the passionate female to apply mouse dung as a liniment, eat snail excrement or pigeon droppings mixed with oil and wine, or rub her loins with blood from the ticks on a wild black bull. That would do it for the male appetite, if nothing else.

Infanticide by exposure was not made illegal in Rome until the 4th century A.D., and it was generally the girls who suffered more, since the ancient "laws of Romulus" ordered families to raise all boys but only the first-born girl. This of course reduced the number of potential mothers. The problem in Rome became too few rather than too many people. Pestilences of smallpox and measles took a large toll, and the stresses of urban life may have resulted in increased miscarriages and sickness. Reay Tannahill observes, too, that the heavy wine intake in Rome was no aid in increasing the population. When not working, the men were either lounging around in very hot baths or getting loaded, both of which can reduce fertility.

The early Christian Church decided that contraception was a sin and, like much of the other learning of the ancient world, scientific knowledge of it was suppressed and lost for many centuries. The heretical woman thus resorted to amulets, superstition, and magic if she wanted no more children and could not ward off her husband's advances.

If women's means of controlling pregnancies throughout history were multifarious, men relied mainly on abstinence or coitus interruptus until the 18th century, when the condom became fashionable. The 16th-century Italian anatomist Gabriel Fallopius is given credit for the invention, initially designed as a protection against syphilis. Two hundred years later, London brothels and various wholesalers carried the contraceptive, and the notorious Casanova endorsed it "to put the fair sex under shelter from all fear."

?

How did they set the price of the Louisiana Purchase?

"Quickly" is the answer to this question—before Napoleon could change his mind.

In the early 19th century, New Orleans was already a vital port, strategically located at the mouth of the long Mississippi. Early settlers in the West shipped nearly a third of their produce in flatboats along its tributaries and down the river to the harbor city for shipment abroad. The right to deposit these cargoes was obtained from Spain, which since 1763 had had a rather feeble hold on the massive territory, stretching west to the Rockies and north beyond the present United States–Canadian border. A shock rippled through the West when news spread that Spain had secretly negotiated the transfer of the territory to France in the Treaty of San Ildefonso (1800). Two years later, the gates of New Orleans suddenly closed on Americans.

President Thomas Jefferson, who was dreaming of America as a quiet, agrarian utopia immune to foreign entanglements, was forced to act. Fearing an outright clash instigated by the irate settlers, he decided to try negotiating with Napoleon. Robert Livingston and James Monroe were dispatched to Paris with $2 million appropriated by Congress for the purchase of New Orleans and western Florida. The two were rather stunned, therefore, when Talleyrand, representing France, proposed the sale of the entire territory. A number of reasons lay behind this move: France was on the brink of war with Great Britain and Napoleon feared the British Navy could block his occupation of Louisiana anyway. And after a slave insurrection in Hispaniola proved victorious against his troops, Napoleon decided to focus his attention on Europe.

After no more than a moment's hesitation, the American diplomats agreed, doubling the size of their country in the blink of an eye. Some haggling occurred over the price, but since the idea was Napoleon's—"It is not only New Orleans that I will cede," he had announced; "it is the whole colony without any reservation"—and since he needed money, the purchase price was modest to say the least: $15 million for the valleys of the entire Mississippi and its western tributaries, an area of 828,000 square miles.

"You have made a noble bargain for yourselves," Talleyrand told the Americans, "and I suppose you will make the most of it."

?

How did they learn the *Titanic* was sinking?

On April 10, 1912, the magnificent *Titanic*, called unsinkable, steamed slowly away from Southampton on her maiden voyage with her cargo of aristocrats, her lavish dining rooms, suites, and cafés. She picked up passengers at Queenstown, then turned her bow westward toward New York, with 1,316 passengers and 891 crew members. Nearly 900 feet long—as long as four city blocks—and 11 stories high, the gigantic ship weighed over 46,000 tons, displaced 66,000 tons of water, and featured a double bottom of 16 watertight compartments.

Popular among the guests was the lively Parisian café, with its wicker furniture and ivy-covered walls. Then, too, one could while away the hours in the ornate Turkish bath, admiring its blue-green mosaic floor and gilded beams. For those who lunched too well at the extravagant buffets of cold salmon, potted shrimp, roast beef, veal, and ham pie, with cold German beer, cheeses, and pastry, there was a gymnasium replete with bicycle machines, or one's cabin and its high-canopied bed. Those with a private suite (for which they had paid $4,350 for the six-day voyage) might stroll on their private promenade deck, lulled by the rhythmic slap of waves, and scan the clear horizon. Such suites were occupied by the cream of New York and Philadelphia society—Colonel and Mrs. John Jacob Astor, Benjamin Guggenheim, the Arthur Ryersons, the John B. Thayers, Harry Widener (son of the streetcar magnate), publisher Henry Sleeper Harper and his prize Pekingese Sun Yat-sen—and they were attended by valets, maids, and nurses over and above the ship's numerous stewards and stewardesses. The voyage, in short, was a fairy tale, a seemingly inviolable domain of leisure and fortune.

On April 14—the fourth day of the voyage—the night was cold, starry, still. The sea's surface gleamed like oil. By 11 P.M. many of the passengers had retired. Off the southeast tip of Greenland, where the *Titanic* knifed along at 22½ knots, the water temperature dropped to 31°. Then came a warning from the nearby *Californian* of icebergs. But the *Titanic*'s weary radio operator, John George Phillips, plagued by a backlog of messages from the passengers, snapped at the *Californian* to shut up and never learned the location of the danger. At 11:40, the ship was at a latitude of 41°46′ N, longitude 50°14′ W, and many on board felt a strange, grinding jolt. To lookout Frederick Fleet, who had seen the towering iceberg less than a minute before and sounded the crow's nest bell, it looked like a close miss. But those down in the engine rooms knew otherwise. Seawater deluged the ship's first six compartments, gushing through a tremendous tear on the starboard side. Sealing off the separate compartments to contain the surge (as the builder had planned for such an unlikely disaster) proved futile. Within 25 minutes Captain Edward Smith ordered the life boats uncovered and officers were dispatched to summon the passengers. At 12:15 A.M. Smith ordered Phillips to send out the regulation distress call: CQD. Six times the signal was emitted. But on the *Californian*, a mere ten miles away, not a sound was heard. Radio operator Cyril F. Evans had closed up shop as usual at 11:30 P.M. Third Officer Charles V. Groves, who frequently toyed with the radio in his spare time, came down that night to listen. He had the headphones on at 12:15, unaware that the radio was inoperative. Bored by the silence, he left and the signal passed on through the freezing night.

The *Californian,* a small boat en route from London to Boston, had stopped that night because of the ice. Various officers on deck had sighted the *Titanic,* and as some of its lights disappeared, they figured the ship was routinely shutting down for the night. In fact, it had veered sharply to port. And later, when all lights vanished, they assumed the *Titanic* had continued safely on its voyage.

Radio operator Phillips, meanwhile, was bewildered by the lack of response from the *Californian*. Captain Smith urged stronger measures. Starting at 12:45 a series of rockets were shot from the deck, searing the night sky with a white light, clearly visible to the officers of the *Californian*. So far was it from anyone's wildest dream that the *Titanic* could be in trouble that the *Californian*'s officers merely thought how odd it all was, and went on about their business. Even those farther-off ships that received the call of distress were slow to fathom the gravity of the situation. At Phillips's desperate insistence, the *Carpathia,* which was 58 miles away and had missed the first message, finally leaped into action. Under the direction of Captain Arthur Rostron the ship changed course and sped at 17 knots toward the sinking *Titanic*. It would be several hours before she arrived, to find only life boats scattered among boulders of ice. But now her wireless operator listened anxiously: "Get your boats ready; going down fast at the head," Phillips told the *Olympic* (500 miles away) at 1:10 A.M. "Sinking head down . . . engine room getting flooded" was heard at 1:35. Cape Race in Newfoundland picked up the news and passed it on farther inland. Even as far away as New York, a wireless operator on the roof of Wanamaker's department store heard the signal and relayed it to an astonished world.

But this ever widening circle of attention was too late, too distant to aid the *Titanic*. The speed with which she went down was staggering. She sank at 2:20 A.M.—not three hours after the collision with the ice—taking hundreds with her. The "unsinkable" ship had boats for only 1,178 of the 2,207 persons on board, and for some inexplicable reason, even these were not filled to capacity. Women and children were loaded first, according to traditional naval chivalry, and as the list of survivors would tell, first-class passengers received first—in many cases the only—consideration. The last boat was put down at 2:05. Many of those left behind leaped into the icy water, clinging to each other and calling in vain to the boats. Others stood on deck, nobly resigned to this unlikely fate. One of these was Benjamin Guggenheim, dressed in his evening clothes to die with the dignity he had possessed in life. Roughly 1,500 lives were lost, as well as untold riches: a jeweled copy of the *Rubáiyát of Omar Khayyám,* Major Peuchen's tin box with $200,000 in bonds and $100,000 in stocks, and the fabulous wardrobe of Baron von Drachstedt. All this still lies at the bottom of the sea in the ghostly tomb that was once a triumph of technology, a model of luxury.

?

How did Bach get the job of cantor at St. Thomas's Church in Leipzig?

Not easily. And not in response to his musical talents.

Johann Sebastian Bach was the third choice of the Council of Leipzig, definitely not the most desirable of the six applicants who turned up in 1722 to compete for the enviable post of music director in Leipzig. Georg Philipp Telemann was indisputably the star. As the prestigious music director and cantor of Hamburg, and a composer and conductor of popular operas, he could throw his weight around; he got the council to drop the requirement that the cantor teach Latin, for instance, then with Leipzig in the bag, demanded more money from his employers at Hamburg. This ploy was successful, and the crestfallen councillors in Leipzig had to turn their eyes elsewhere, reluctantly, skimming over Bach (employed at the tiny court of Cöthen) to their second choice, Christoph Graupner. His credentials were nearly as glowing as Telemann's, and the council eagerly offered the position to this

Johann Sebastian Bach's expertise at the keyboard and genius for composition were of little import to the employers of his day.

conductor of Prince Hesse's orchestra in Darmstadt. Graupner was ready to pack his bags when Prince Hesse staunchly refused to let him go, and softened the blow with a raise and a bonus.

The distraught councillors, now over six months without a cantor, turned to Bach. He was a well-regarded organist, but that didn't impress them. They had heard only two of his compositions (two Mühlhausen cantatas) for that was all he had in print at the time. Bach's present position in Cöthen was unspectacular and, worst of all, he hadn't been to college.

The point that ultimately tipped the balance in Bach's favor was his concession to teach Latin and Latin catechism in St. Thomas's school. When the results of an election on April 22, 1723, were tabulated, the council had agreed on a new cantor—first and foremost, a Latin teacher; second, one of the most prolific and inspired geniuses in the history of music.

?

How did they excavate Delphi?

Hovering over the steep Pleistus Valley and shielded by rugged Mount Parnassus is the ancient sanctuary of Apollo at Delphi. Here at the crossroads of several trade routes, men of the mercantile world would meet and exchange wares. Others would come to seek the wisdom of the fabled oracle, the frenzied prophetess through whom Apollo spoke. Today the white stone ruins gleam in the Mediterranean sun, an elegant theater lies nestled in the hillside, and on the hilltop overlooking rocky slopes and the Gulf of Corinth is a spacious stadium with gray stone seats. The site is one of Greece's finest and most famous, but for centuries it lay buried, its extent unknown.

When the Danish archaeologist Peter Oluf Brøndsted visited Delphi in 1810 he encountered a major obstacle: "The wretched little village of Kastri in many ways renders it difficult to survey the whole site, and to be able to get a satisfactory plan of Delphi one would have to begin by pulling down many of its huts." The

Greeks of this village did not find it so wretched, however, and were in no mood to raze their homes and pull out their crops in the name of scholarly research—particularly a foreigner's—so some years went by without progress.

The French entered the scene in 1861. Backed by Napoleon III, Paul Foucart and Karl Wescher executed the first regular excavation, detecting the eastern limit of the temple precinct. But when King Otho was expelled from Greece, all work on the site was halted for nearly 20 years.

In 1880 a French scholar, Bernard Haussoullier, uncovered the Stoa of the Athenians and a segment of the processional road. Foucart, then director of the French School at Athens, applied to his government for 100,000 francs and established an agreement with the Greek government by which it would expropriate 30 houses in Kastri. A change of government again snarled the proceedings. The new minister would keep to the plan only if the French agreed to lower the duty on Greek currants. When the French refused, the Greeks offered the opportunity of excavating to the Germans, who politely declined out of goodwill toward the French. Over the next decade the newly founded Archaeological Institute of America launched an enthusiastic campaign to raise funds and gather support so that it might do the job, but then the French concurred on the notorious currants.

In 1891 a definitive contract was signed, giving the French exclusive rights to excavate the site during the next ten years. The French government voted a grant of 500,000 francs for the endeavor, which caused one outraged Marquis in the Senate to exclaim, "*On fait de nous des chercheurs de truffes!*" The Greek government, supplying 60,000 drachmas, agreed to move 100 houses a short distance to the west.

Although the governments may have been conciliatory, the uprooted Kastriotes were not. When work began in 1892, the villagers attacked the workmen and seized their tools, wreaking such havoc that the Greek government was forced to call in the military. Thereafter a ring of soldiers with loaded rifles surrounded the scholars, who quietly dug, examined, analyzed, and recorded, shielded from the present in their search for the past.

For seven consecutive summers the archaeologists worked under the superb guidance of Théophile Homolle, by then director of the French school. One of the earliest finds was the Treasury of the Athenians. Homolle in great excitement cabled Paris, and the Greek authorities in Amphissa, the capital of the district, came running—but their ardor was hardly intellectual. ". . . A slight misunderstanding arose," explained Homolle, "and in their artless way they [the Greek government] hoped that ready money had been found in the ground, at an opportune moment, to pay off their interest due."

By 1900 much of the 2,400-year-old site had been uncovered. Since then hordes of scholars, students, and tourists have traveled to the lofty slope each year, marveled at the spare Doric columns, stood before the *Charioteer*, imagined the cryptic cries of the priestess, and perhaps spent a little money in town—the sole compensation for those who once lived on the sacred place.

?

How did Charlie Chaplin make up his tramp costume?

Chaplin's trademark—the derby cocked on his head, thick swatch of a moustache, and gargantuan shoes—was the result of Mack Sennett's request that he "get into a comedy make-up" for *Kid Auto Races at Venice* (1914). With a little ingenuity, he concocted the outfit in a dressing room while Fatty Arbuckle and Chester Conklin played pinochle. Fatty kindly contributed his father-in-law's derby and his own pants, of generous proportions. Cocklin provided the little cutaway and Ford Sterling the size-14 shoes, which were so big Chaplin had to wear each on the wrong foot to keep them on. He devised the moustache from a bit of crepe hair belonging to Mack Swain. The only thing the tramp himself owned was the whangee cane.

?

How did the United States buy Alaska?

At 4 A.M. on March 30, 1867, Secretary of State William H. Seward and Baron Edouard Stoeckl, Russian Minister to the United States, signed the Treaty of Cession, by which Alaska passed from Russian to American hands for a mere $7.2 million, culminating three weeks of intense negotiation carried on in unusual secrecy. Seward rushed the treaty to a secret session of the Senate for immediate ratification, fearing that delay would allow time for opposition to mount. He enlisted the support of the influential Senator Charles Sumner, who became a champion of the cause but privately admitted, "Abstractly I am against further accessions of territory but this question was so perplexed by considerations of politics and comity and the engagements already entered into by the government, I hesitated to take the responsibility of defeating it." This spokesman, however lukewarm on the subject, presented a three-hour speech to the Senate in which he stressed the riches in furs and fish to be reaped from the new land, and the obligation of the United States to act favorably toward Russia, which had supported the Union (where England and France had not) in the recent Civil War. Sumner also stressed the creation of new industrial interests on the Pacific and unlimited commerce with Japan and China, which would contribute to the supremacy of the United States throughout the world.

It was this argument and all that it implied that was most closely related to Seward's own ambitions and notions of grandeur. Drawn to both the power of manifest destiny and the drama of a limitless American frontier, Seward saw himself as leader of jingoist Americans impatient to see all of North America under United States dominion. ". . . The Russian outposts in North America," said Seward stoutly, "will yet become the outposts of my own country." With six members absent from the Senate, the treaty was narrowly ratified by a vote of 27 to 12 (a two-thirds majority was required). Another vote was taken in an effort to make it unanimous, resulting in a final tally of 37 to 2, which constitutes the official record today.

News of the acquisition was released to the press in early April 1867. Then began the diatribes against Seward and his "Icebergia" or "Polaria," which, according to the *New York Herald*, was "an ice house, a worthless desert with which to enable the Secretary of State to cover up the mortification and defeats he has suffered with the shipwrecked Southern policy of Andrew Jackson. . . ." The *New York Tribune* said, "Ninety-nine hundredths of Russian America are absolutely useless; the remaining hundredth may be of some value, but is certainly not worth seven million dollars. . . . The expense and trouble of a territorial government . . . in this distant and uninhabitable land would far outweigh any advantage from its codfish or bearskins. . . . By the next session of Congress we trust the folly of the purchase will be made so plain that the House will refuse to make the necessary appropriation." Thus was the myth of "Seward's Folly" foisted upon Americans, featuring Seward as the sole proponent of a foolish venture that only later by some mysterious act of providence became a heroic coup of inestimable value.

The entire subject of Alaska's acquisition is, in fact, riddled with ambiguity, tainted with bribery and tangled political intrigue. Seward was not the sole champion of the cause, nor the first. Nor were the Russians laughable fools, as has been suggested, with no understanding of what they had on their hands.

Russia's occupation of the land that is now Alaska lasted 126 years, significant mainly in that it prevented Great Britain from assuming what would probably have been lasting control. Russia's interests were exclusively economic and her presence was embodied in the Russian-American Company, which held a monopoly on trading rights as far south as the 55th parallel. The Grand Duke Constantine, brother of Czar Nicholas I and head of the Admiralty, noted designs of America and Great Britain on this province and recognized Russia's inability and unwillingness to defend it if the need arose (preferring at the time to consolidate troops on the mainland). The powerful Hudson Bay Company had wangled a lease on southeastern Alaska in return for 2,000 otter skins annually (a lease that conveniently expired in 1867); the British had established a trading post at Fort Yukon; American whaling boats were encroaching on Russian waters; 1857 brought rumors of a planned Mormon colony on Alaskan soil; and as early as 1832 Russia knew about the presence of gold in Alaska but concealed it in fear of an onslaught of hungry prospectors. Thus, Constantine anticipated economic infiltration and loss from within, and advocated sale of the province before this should occur. Russia, furthermore, had few friends in the world and wished to maintain friendly relations with America, particularly in union against a traditional foe, Great Britain. Senator William Gwin early recognized Alaska's economic potentialities and, speaking unofficially for President Buchanan, offered Stoeckl $5 million for the purchase in 1859. The Czar and Foreign Minister Gorchakov, originally opposed to the idea, began to entertain the proposal seriously, but the Civil War intervened and negotiations stopped.

During the war years an enterprise was begun that, although it ultimately failed, maintained American contact with Russia, interest in Alaska, and power in the Pacific and in Asia. Ambitious entrepreneur Perry McDonough Collins founded the Collins Overland Line (a subsidiary of Western Union), whose goal was to lay a cable from California north through British Columbia, Alaska, and the Bering Strait to the mouth of the Amur River in eastern Asia. His imperialistic attitude, quite popular at the time, is evidenced in his remark "The earth will persist in being eight thousand miles in diameter and some twenty-four thousand miles in circumference. [As] everybody does not live on Manhattan Island, we very confidently conform to circumstances, and take the world and whatever there is in it, or upon it and do the best we can." Collins and Hiram Sibley, president of Western Union Telegraph Company, met with Gorchakov in 1864 and learned of the Foreign Minister's willingness to sell Alaska. At the same time, U.S. Minister to Russia Cassius Clay advised Seward of the benefits to be derived from the purchase.

In 1866 Constantine, with permission from the Czar, sent Stoeckl to Washington to close the deal for an amount no lower than $5 million. The price ultimately agreed upon in the treaty was $7.2 million. Apparently only about $7 million ever reached Russia and this after a bitter fight in the House, its members slighted and outraged by Seward's stealthy handling of the deal and the Senate's hasty ratification. The bill was in fact presented for appropriation months after the transfer of territory had occurred at Sitka, delayed partly by impeachment proceedings against President Andrew Johnson. Perhaps no one really knows where the balance of the money went (although a House investigation was undertaken to find out), but rumors were rampant of bribery both of the press and politicians, including General N. P. Banks, Chairman of the Committee on Foreign Affairs Thaddeus Stevens, and the brazenly unscrupulous Robert J. Walker, a former senator. Seward and Stoeckl, both eager for feathers in their caps, feared the opposition of ardent anti-expansionists, with good cause. Members of the House hotly proclaimed that the land was "virtually valueless" and an irresponsible waste of money. ". . . To suppose that anyone would leave the United States," said Representative Benjamin F. Loan, ". . . to seek a home . . . in the regions of perpetual snow, is simply to suppose such a person insane." Together Seward and Stoeckl enlisted the aid of Walker, who claimed to have urged President Polk to acquire Alaska as early as 1845. Walker wrote a series of articles endorsing the purchase, which appeared in the *Washington Daily Morning Chronicle,* whose editor, John W. Forney, received $30,000 in return. Walker, himself amply paid, took the liberty of invading the House of Representatives to stress the attractions of Alaska—the fish, gold, minerals, even the "charming" climate of Sitka. According to the *New York Herald,* the "illustrious Premier," Seward, was meanwhile "working the telegraphs and the Associated Press in the manufacture of public opinion night and day. . . ."

On July 14, 1868, the bill was passed in the House by a vote of 113 to 43, with 44 abstaining. The complacent Walker went to New York with $21,000 in gold certificates and $2,300 in greenbacks. There, his pockets were immediately picked of $16,000 worth of gold Treasury certificates! The police managed to find the thieves but Walker, because of the delicacy of the situation, declined to press charges—just one of the suspicious circumstances that resulted in an investigation of the funds.

Alaska's entry into the United States certainly had a dubious and bungling beginning, signifying no long-range policy on the part of the government. Stoeckl's readiness to grease palms and Seward's persistence and unwavering enthusiasm took center stage, but the maneuver involved a wider cast of characters and more developments than are generally remembered. Seward was not alone in his "folly"—many newspapers across the country supported him—but even his advocates could anticipate a mere fraction of the tremendous riches that would eventually come from "the region of perpetual snow."

?

How did John Quincy Adams meet the press?

John Quincy Adams, who served as President from 1824 to 1828, was noted for his austere reserve and cold manner. Critics labeled him an aristocrat and tyrant. Even Anne Royall, the caustic newspaperwoman who left no scandal concealed, no political machine unscathed, found Adams a difficult man to get to. Time and again she pounded on the White House door, eager to interview the President on the controversial topic of state banks. Frustrated in her attempts, "the Mother of Yellow Journalism" resorted to sheer doggedness. She spied on the President's daily routine and discovered that he ventured down to the Potomac in the dim light of dawn and there shed his dressing gown and slippers to plunge stark naked into the balmy water.

As Adams paddled toward shore one sunny July morning, he was assailed by a peremptory voice—"Come here!" There, primly seated on his clothes was Anne Royall, whom he knew well. The elderly reporter announced her intentions. ". . . I stalked you from the Mansion down here. I'm sitting on your clothes and you don't get them till I get the interview! Will you give it to me or do you want to stay in there the rest of your life?"

The President now stood in water up to his chin, due modesty restraining him from advancing farther. "Let me get out and dress," he beseeched her, "and I'll promise to give you the interview. Please go behind those bushes while I make my toilet."

"No, you don't!" snapped Royall. "You are President of the United States and there are a good many millions of people who want to know and ought to know your opinion on this Bank question. I'm going to get it. If you try to get out and get your clothes

I'll scream and I just saw three fishermen around the bend. You don't get out ahead of that interview!"

Not wishing to broadcast the already delicate situation, the President conceded. He stood helplessly in the water, spluttering and squirming beneath a barrage of questions, while Anne enjoyed a position that most journalists who cover press conferences today would kill for.

?

How did Howard Hughes find his women?

With the demise of his first marriage in the early '20s, Hughes abandoned himself to 30 years of rampant promiscuity—bizarre and often underhanded. Enamored of Hollywood, he was equally smitten by its leading ladies. Marian Marsh, Jean Harlow, Ava Gardner, Lana Turner, Ida Lupino, and Carole Lombard were all associated with the tall and mysterious Texan. Hughes took a liking to Katharine Hepburn and once followed a Hepburn show across the country in his private plane—all in vain, for Kate pronounced him a bore.

Hughes's appetite proved insatiable, and since he could never maintain a permanent relationship with a woman, he was forever on the prowl—and not only he, but also his numerous operatives throughout the country who amassed files on likely candidates for Hughes to peruse. Many of these (primarily shapely brunettes) were swept off to Los Angeles, assured a career in the movies, issued a salary, and kept in one of Hughes's various pads for his own convenience. Needless to say, the careers did not materialize.

Hughes was an avid viewer of beauty contests. In 1960 he selected his own seven favorites from the Miss Universe pageant, and had his detectives lure them to hotels in Los Angeles where they were to await their rise to fame. These young women caught on to the ruse, however, and fled.

Unlimited wealth added some astounding stratagems to

Hughes's courtship procedures. Having seen a picture of the Ballet de Paris prima ballerina, Zizi Jeanmaire, Hughes became infatuated and sent a studio executive, William Fadiman, to see her and confirm the grounds for his attachment. Fadiman was taken with the dancer, so Hughes promptly bought the entire company and moved it to the RKO Writers Building in Los Angeles. He planned a movie and the seduction of Zizi, but when the latter failed after two years, he abandoned the former as well.

Hughes encountered obstacles in yet another of his escapades. As he was making love to a 16-year-old blonde with a 40-inch bust, the girl's mother burst in and caught them in the act. It cost the billionaire some $250,000 to avoid charges and keep the situation out of court.

Hughes's obsessive sex life ended, as was fitting for the eccentric, in a peculiar way. He became paranoid about germs and came to dread physical contact with anyone. However, rather than save the money he had once expended on his libido, he used it to feed a severe addiction to codeine and Valium.

?

How did they know there was an El Dorado?

Nestled in the mountains surrounding Bogotá, Colombia, lies a rather unspectacular lake called Guatavitá. But a very strange ritual was once performed here, one that caused Spaniards and Englishmen alike to forsake the comforts of home, battle alligators and jaguars and cannibals, suffer scurvy and starvation, and lose their way in the torrid jungles—all for the allure of gold. According to an ancient legend, this little lake was formed thousands of years ago when a meteor struck the earth. The memory of this bizarre event was passed down through the generations in the form of a myth: that one day long ago a golden god dropped from the sky and settled in the bottom of Lake Guatavitá.

The Muisca Indians of the Chibcha tribes lived and farmed near the sacred lake, and with the election of each new chieftain (or on other designated days), they gave thanks to the god who inhabited it. They rubbed the new leader with balsam gum and then dusted him with gold, sprayed through cane tubes. What a dazzling and strange sight it must have been—this golden statue striding toward the lake accompanied by priests in black robes and Indians painted red. He took a raft to the center of the lake and dove in, while people on the banks threw emeralds and gold objects to please the god in the waters. Evan S. Connell reports in *A Long Desire* that the last performance of this ritual was in 1480, but talk of it only increased with time, and fantastical accounts spread throughout the region and across the seas. At length, the ritual became fused with the dreams of avaricious Europeans, eager for plunder, power, and reward. They decided this gilded man was the omnipotent ruler of a golden empire, where he and his nobles dressed in gold each day and washed it off at night, where soldiers wore golden armor and buildings were coated in gold.

A German by the name of Ambrosius Dalfinger was the first to seek El Dorado, the golden man with his lustrous city. Shirking his duties as Governor of Venezuela, Dalfinger plunged into the jungle in 1529 with 200 soldiers. He managed to obtain some gold objects, not in vast quantities, but the quality of craftsmanship surpassed the abilities of the Venezuelan Indians. They came, he learned, from a wonderful place called Xerirá, where emeralds and gold were abundant and easily bought. This was enough to keep Dalfinger searching for the next ten months, but in vain. After tending to some business at home, he set out again in 1531, this time convinced that all the Indians knew the location of Xerirá and, if sufficiently tortured and terrorized, would relinquish the information. His horrific deeds resulted in the collection of 100 kilos of gold (which was dispatched to his home but lost in transit in the jungle) but no golden kingdom. Two years after setting out, survivors of the hapless mission straggled back to their town of Coro with tales of vampire bats, yellow fever, syphilis, and poisoned arrows, one of which had found its mark in Dalfinger's throat.

A lawyer disenchanted with his profession was next to pursue the city of gold. In 1536 Gonzalo Jiménez de Quesada started upriver from Santa Marta on the Colombian coast, purportedly to find a quick passage to the South Sea for Charles V of Spain. But he and his 900 men had gold rather than spices in mind, and were willing to suffer every pestilence imaginable to attain it. Cannibals captured and ate some; wild beasts feasted on others; the remainder were racked by disease and starvation. With nothing to eat but insects and bats, the hides of dead horses, leather shields, and raw corn, those who were able trudged on. After a year, with 600 dead, Quesada reached the lofty plateau of Cudinamarca, home of the Chibcha Indians. He easily seized the town of Tunja at the foot, where the houses were decorated with thin sheets of gold. Indians showed Quesada Lake Guatavitá, but somehow this didn't jibe with the Spaniard's expectations. Nor did the Indians' insistence that the gilded chieftains were a thing of the past convince him. Quesada stayed on the plateau for two fruitless years in search of an empire of gold.

Just as Quesada was contemplating a return to Spain to obtain governorship of this territory he called New Granada, an uncanny coincidence occurred. Two other armies, each unaware of the existence of the other and of Quesada's, converged on the Cudinamarca plateau, the reputed kingdom of gold. From the southwest came the lavishly dressed Governor of what is now Ecuador: Sebastián de Belalcázar, a captain of Pizarro, with his well-armed soldiers and an ample supply of pigs, cattle, and horses. At nearly the same time, a ragged, feverish, and very hungry troop trickled in from the northeast; this was led by Nikolaus Federmann, who had marched 1,000 miles in search of El Dorado. Quesada's gifts of emeralds and gold proved sufficient to placate the advancing armies, and the Spaniard, by now well accepted by the natives, retained his position, formalized in a ceremony assigning all lands to Charles V. Still, despite the concurrent arrival of the gold seekers and the resultant hubbub, the elusive city did not appear. And the aftermath of the adventure was a definite anticlimax: Federmann was accused of theft by his German employers and died in 1541; Quesada, still wishing to be Governor of New Granada, lost out to his financier's son because of some intrigue at court. The hardy conquistador continued to dream of El Dorado and died penniless in 1579, a victim of leprosy. Belalcázar fared a little better, returning to govern the northern part of the Inca realm; but after killing a challenger, he was recalled to Spain and died en route.

Quesada's younger brother, Hernán, rekindled the fires of the El Dorado dream after the three conquistadores had departed from Guatavitá. He engaged some Indians to start emptying the lake so that he could find the gold trinkets that, according to legend, had been tossed there. Dipping calabashes into the edge of the lake, the Indians passed them along a line of workers stretching to a surrounding hill. Then the gourds were emptied and sent back. After several months, Hernán took a good look at the lake bed and found a few objects—not gold, however, but gilded copper. The heavier gold objects must have slipped into the middle of the lake. Now Hernán had tangible evidence of the veracity of the legend, but this tedious search was decidedly unromantic, and he gave it up. Instead he embarked on a disastrous search *elsewhere* for the shadowy El Dorado. He forfeited the lives of thousands of Indians and over 100 Spaniards—all for naught—and was later struck by lightning.

Undoubtedly one of the most persistent pursuers of El Dorado was Quesada's nephew, Antonio de Berrio, a captain of the guards in Granada. He happened to inherit his uncle's estate in the New World, which in 1580 was producing a comfortable income. But if he was to have it, read Quesada's will, he would have to carry on the search for El Dorado. Rather reluctantly, Berrio read up on the previous star-crossed expeditions, but the horror, the mystery, and the romance took hold. For 15 years, Berrio's wife and daughters sat at home while he traipsed through the sultry jungles, meeting and talking to everyone. One such acquaintance was Juan Martín de Albújar, a survivor of an expedition ten years before. This wild character offered to swear on the Bible that while a captive of the Caribs, he had been led blindfolded through the jungle to a vast city called Manoa, by a Lake Parima. There, free to see, he walked and walked past stone houses to a glorious palace where lived the Gilded Man. He stayed seven months, telling the luminous figure about Spanish customs and then left (again blindfolded) with gifts of gold and jewels, which he proffered as evidence of his tale. This ignited Don Antonio to try once more, but at that point in 1595, he met conflict and competition with the encroaching English.

Sir Walter Raleigh was Don Antonio's bête noire. He seized Berrio and pumped him for information, of which he got little, for Antonio pretended to be senile. Raleigh headed up one of the mouths of the Orinoco and promptly got lost. He took a gentler approach to the Indians than had the Spaniards, and was continually reassured that indeed just beyond the mountains was a kingdom of gold. At a point 300 miles upriver, Raleigh decided to turn back, but left two men to carry on his search. Hugh Goodwin was to learn the language of the Indians; Frances Sparrey was to march on to the golden city, study its fortifications, prepare methods to seize it, and return to the Orinoco. There he might meet Raleigh again, or go on to the coast and hitch a ride with another English ship. But the unlikely plan collapsed, for the Spaniards caught Sparrey and threw him in jail.

In 1603, Sir Walter Raleigh as well found himself imprisoned, in the Tower of London, charged with having failed to report plots of treason. Since his return to England, he had countered his disappointed and skeptical sponsors by writing a huge and rather fantastical opus titled (in brief) *The Discoverie of the Large, Rich, and Bewtiful Empyre of Guiana, with a Relation of the Great and Golden Citie of Manoa (which the Spanyards call El Dorado).* Here Raleigh recounted the ancient ritual of the gilded man, and Lake Parima, reportedly lying between the Orinoco and the Amazon, suddenly popped up on all maps of South America and remained there until 1800. Raleigh was sentenced to die, but in 1616 he stepped out of the Tower toward freedom, only because of some persuasive talk about finding El Dorado for the glory and purse of King James. His flagship, the *Destiny,* set sail in 1617, but once again it was the white man's fate to suffer for greed. Forty-two men died by the time the ship reached South America, and Raleigh contracted malaria. He sent a group upriver; it was devastated by Spaniards. Those who survived turned tail and sailed back to England. Raleigh had only managed to extend his life two miserable years. In October 1618, again accused of treason and considered a pirate, he walked to the scaffold and there spoke his final words, an eloquent speech of 30 minutes.

Today a hint of the mythical El Dorado may be seen in the Museo del Oro of the Banco de la República in Bogotá. Nearly 10,000 pre-Columbian gold artifacts reside here, including brooches, masks, spoons, nose rings, and images of men and beasts, many of which were made by those quiet farming folk the Muisca.

?

How did they kill Trotsky?

The first attack came on May 24, 1940. Leon Trotsky, the powerful revolutionary and political strategist, Lenin's comrade and appointed successor, had been living in exile for 11 years. Though his health was failing, he continued to work feverishly from behind the high walls of his villa in Coyoacán, Mexico, writing and plotting worldwide revolution. At 4 A.M. his wife Natalie tried urgently to wake him; he was dazed from a sleeping powder taken the night before. Within seconds the two were on the floor, pressed tightly against the wall, as a stream of bullets tore up their bed. In the doorway were silhouettes of helmeted figures with Thompson submachine guns firing wildly throughout the room, echoing shots from the room of Trotsky's grandson. The raiders hastily departed, leaving the three, miraculously, with only minor wounds. An immediate search revealed a time bomb loaded with three pounds of dynamite, which had failed to go off, several Mausers, an iron bar, scores of submachine-gun cartridges, and assorted clothing, all left behind by the would-be assassins.

For some time the Mexican Presidential Police maintained that Trotsky must have rigged the entire scene, since it was so unlikely he and his wife could have escaped the rain of 200 bullets. Nevertheless, the police chief asked Trotsky who he thought was responsible. With uncanny calm and drama, Trotsky led the chief into his garden and whispered in his ear, "Joseph Stalin!"—his avowed enemy, the man who had outmaneuvered Trotsky in the succession to Lenin.

In the following days, security was tightened. The number of police stationed around Trotsky's villa was increased from 5 to 15, in addition to the 4 young American Trotskyites who stood guard. Steel shutters were attached to the windows, the doors lined with steel, the roof reinforced with concrete. All of this was noted with considerable interest by one Jacson-Mornard, a young man seen frequently around the fortress, known only as the lover of Sylvia

Ageloff, one of Trotsky's secretaries. He approved of the defensive measures, Mornard told Trotsky and his guards, adding, "Of course, in the next attack the GPU [Soviet political police] will use other methods."

The mysterious suitor managed to insinuate his way into the house and Trotsky's presence more than once. Politely offering to take some American Trotskyites on a sight-seeing tour, he deliberately swerved on a precipitous mountain road, nearly making a violent end to their plans. The handful of delegates from Minneapolis were meeting with Trotsky to discuss nothing less than a revolution in America, to be launched in their hometown. Mornard heard a great deal.

On August 20, 1940, a blazing summer day, Mornard drove up to the fortress and parked his Buick facing Mexico City, rather than facing the wall as was customary. Three of the guards (who failed to notice the difference) were busily installing a siren on the watchtower as Mornard blithely waved to them, indicating he was there to see Sylvia. They let him through the steel doors, where he passed the fourth guard without the slightest questioning. Inside, Natalie greeted the young man, commenting immediately upon the overcoat and hat Mornard carried over his left arm, clutched tightly to his body, despite the heat. Mornard insisted it was going to rain. After chatting idly with Trotsky for a few minutes in the garden, the two men retired to Trotsky's study to discuss an article that Mornard was writing. Trotsky was not at all interested, having seen a first draft in his study with Mornard present two days before, but for some reason he acquiesced to the young man's wishes.

Jacson-Mornard, wretchedly pale and obviously overwrought, laid his coat over Trotsky's desk, carefully obstructing the switch that activated an alarm system. Within moments he seized his weapon. The coat hid a dagger, a revolver, and an ice pick, and Mornard lighted upon the last, for silence was essential. He leaped at Trotsky, foolishly striking with the broader end of the instrument. Although he succeeded in plunging the ice pick several inches into the side of Trotsky's head, his victim retaliated. As the two wrestled, Trotsky's glasses were smashed. Finally, he let out a blood-chilling scream that brought the guards and Natalie on the run. The guards grappled with Mornard, grabbing his revolver and pinning him to the ground. Trotsky meanwhile staggered outside and collapsed on the patio, blood pouring from his wound, as Natalie tried to comfort him. After about ten minutes, Mornard was again struggling and Trotsky, still lucid and surprisingly calm, ordered that he not be shot since he must talk.

Due to the confusion and the length of time it took to subdue Mornard, some 30 minutes passed before a doctor appeared, directly followed by an ambulance. By the time Trotsky reached the hospital he was in a coma. An extensive trepanning operation on his skull proved useless, and Trotsky died at 7:25 P.M. the following day. His ashes were buried in the garden of his Mexican fortress. A legacy of 20 books survives for the Western world, but in the Soviet Union his name has been obliterated, his deeds and ambitions dropped from the annals of history.

It was years before the true identity of the assassin was revealed; his motives remain obscure to this day. For six hours every day for six months, a team of psychologists questioned Jacson-Mornard in prison in Mexico City. Their exhaustive dossiers, amounting to over 1,000 pages, revealed nothing conclusive. A long letter found on his person, dated the day of the murder, strove to justify the act as the outrage and vengeance of a former Trotskyite, severely disillusioned by the opportunistic, selfish motives he saw in the leader. The letter even dragged Sylvia into the picture, asserting that since Trotsky disapproved of her, its author was carrying out the deed for her as well, thus sacrificing himself for his love. Police and psychologists quickly ascertained that the letter was the work of many. And as Robert Payne suggests in his biography of Trotsky, the letter makes some sense only if one considers its mirror image—that is, the Trotsky in the letter actually stands for the powerful GPU agents, who made Mornard their tool, who were themselves corrupt and self-serving, and who undoubtedly disapproved of Sylvia.

Jacson-Mornard, who persisted in baffling everyone, was sentenced in 1943 to 20 years in prison. Seven years later one of the psychologists, Dr. Alfonso Quiroz Cuarón, traveled to Spain with a set of Mornard's fingerprints, and at last the Madrid police solved the mystery. The assassin was Jaime Ramón Mercader del Río, arrested as a communist organizer in 1935. Pardoned a year later, Mercader became a lieutenant in the Republican Army, from which he was recruited by the Soviet general Kotov. While Mercader was trained by Soviet intelligence to work with the GPU, his mother, a wild, suicidal woman prone to ranting about communism, became the lover of General Kotov. Soon Mercader was traveling the world with a fake Canadian passport, for the glory of Stalin and the GPU.

Ramón Mercader was released from prison in 1960. He was given a diplomatic passport with the name Jacques Vandendreschd and swiftly whisked away behind the Iron Curtain. Flown first to Havana, then to Prague, never again seen in the West, he remains a shadowy and rather pathetic figure, whose entire story is yet to be unraveled.

?

How did they decide how tall to make the Empire State Building?

On October 1, 1929, a wrecking crew of 700 arrived at the stately Waldorf-Astoria Hotel to begin demolition of what had been a fashionable center of New York social life for 36 years. On this site, imbued with tradition, several leaders of big business would build the tallest skyscraper to date, one surpassing others in beauty as well as height, one that would meet the demands of all who occupied it—or so the entrepreneurs claimed. In the forefront of the enterprise was John J. Raskob, a principal in General Motors, who early realized the site would become a thriving business center. He was joined by Pierre S. du Pont, president of E. I. du Pont de Nemours, Louis G. Kaufman, and Ellis P. Earle. To assume prominence in the public eye, they selected the charismatic Alfred E. Smith, ex-Governor of New York, as president of Empire State, Inc.

Initially, Raskob thought of building a relatively low, stocky structure of only 30 stories. But like many others, he became enthralled with the race toward the skies that New York had wit-

The Empire State Building in 1930 begins its race toward the skies.

nessed over the previous 20 years. The Metropolitan Life Insurance Company's tower, at 700 feet, held the record until Frank Woolworth, angered by Metropolitan's refusal to give him a loan, in 1913 bested the building by nearly 100 feet with his cathedral on City Hall Park. During the '20s the Bank of Manhattan leaped ahead, rising to 927 feet, and the Chrysler Building opened in 1930 with a surprise finish: a secret spire rose from within the building to a height of 1,046 feet.

In 1929 John Raskob met with architect William F. Lamb of Shreve, Lamb, and Harmon. According to Hamilton Weber, rental agent for the new building, "John J. finally reached into a drawer and pulled out one of those big fat pencils schoolchildren liked to use. He held it up and he said to Bill Lamb: 'Bill, how high can you make it so that it won't fall down?" Eighty stories was the reply, duly announced to the public that summer—enough stories to make it the tallest building in the world.

Other considerations heavily influenced the size and appearance of the building. Its mass was dictated by its budget; weighing the extent of the investment against cost per cubic foot, the corporation arrived at a total size of 36 million cubic feet. (Total construction costs, cut severely by the Depression, came to only $24.7 million.) The design, too, was limited. New York's strict zoning laws required that tall buildings get narrower as they rose. Only a quarter of the proposed building, which would stand on a site of two acres, could rise to any substantial height. Time was a factor as well. The entrepreneurs insisted the building be completed by April 1931, a mere year after construction began, because May 1 was the date many office leases expired in the city. (Amazingly enough, this demand was met, with thousands of workers putting in a total of 7 million man-hours of labor.) Elevators were a factor in the height, particularly since the much narrower upper floors offered limited space for the large number of elevator banks required.

These strictures were juggled around, analyzed by experts, and eventually incorporated into the design of William Lamb, who went back to the drawing board some 15 times before everyone was satisfied. Furthermore, as they watched the Chrysler Building go up, the builders realized even 80 stories wouldn't do, so they upped the figure to 86.

Intrepid workers take in the view from the towering Empire State Building, which surpassed the 1,046-foot Chrysler Building, seen in the background.

Still, Raskob was nervous. His building would have a slender margin of only four feet over the Chrysler building. "What [it] needs is a hat!" exclaimed Raskob, and he concocted the bizarre plan of erecting a huge mast to crown the building and serve as a mooring for dirigibles. ". . . The directors of the Empire State, Inc.," announced Smith, "believe that in a comparatively short time the Zeppelin airships will establish trans-Atlantic, trans-continental and trans-Pacific lines, and possibly a route to South America from the port of New York. Building with an eye to the future, it has been determined to erect this mooring tower to land people directly on Thirty-fourth Street and Fifth Avenue after their ocean trip, seven minutes after the airship connects with the mast." Thus a 200-foot mast was constructed and, in September 1931, a Navy blimp attempted to dock. A violent updraft caused some problems, however, tilting the blimp and upsetting its water ballast, which poured down on unwitting pedestrians blocks away.

The mast idea was discarded, and the mooring grew fatter and became a tower. Its principal purpose was eventually realized, though, for the building now soared to a dizzying height of 1,250 feet, well beyond any other on Manhattan's crowded skyline. A television tower erected in 1950 added yet another 222 feet, and the building enjoyed 39 years as the world's tallest.

?

How did they invent the potato chip?

Potato chips were originally called Saratoga chips after the site of their discovery. In the late 19th century a Native American named George Crumb worked as chef for Moon's Lake House in Saratoga Springs, New York, where a fashionable crowd convened to take the waters of the spa. A persnickety guest reportedly disliked the cut of his French fries one night, and kept sending the oversized potato strips back to the kitchen for a more refined look. Crumb, finally exasperated by the guest's unreasonable persistence, decided to cut the potatoes just as skinny as he could. He boiled the slices in fat and presented them to the complaining diner, who was delighted and didn't think twice about the indecorous crunching and lip smacking and greasy fingers that accompanied their consumption. From this elegant dining room, the new and scarcely wholesome potatoes traveled throughout the nation, becoming one of the largest-selling snack foods in America.

?

How did Dylan Thomas die?

At about 3:30 A.M. on November 4, 1953, Dylan Thomas staggered from the White Horse Tavern to his hotel room in the Chelsea section of New York and announced to his girl friend, Liz Reitell, "I've had eighteen straight whiskeys. I think that's the record." That, at any rate, became the legend. Today students, writers, aspiring poets, and idly curious tourists visit the quaint White Horse Tavern on Hudson Street, Thomas's favorite haunt in New York, with its old wood floors and lanterns, its gleaming oak and mahogany bar with a brass rail and mirrors behind, reminiscent of an English pub. "This is where Dylan Thomas hung out," they muse. "He drank eighteen whiskeys straight and fell off that stool there—dead."

Thomas was a man of violent emotions and sometimes outrageous behavior, a man whose reputation easily lends itself to a dramatic, even romantic end. Perhaps this is why the actual sequence of events preceding his death has been distorted, and a myth has grown up about both the poet and the old tavern. According to his biographer Paul Ferris, Thomas did not die in the White Horse, nor upon reaching his room in Chelsea. He died at St. Vincent's, a private Roman Catholic hospital, after lying in a coma for five days following the fateful night. The cause: alcoholism and pneumonia with pial edema, or pressure on the brain. What has never been conclusively shown, however, is what effect a rather large dose of morphine, administered on November 4 by Dr. Milton D. Feltenstein, had on Thomas. In the past, this doctor had not been reluctant to treat Thomas with injections of ACTH and the poet had referred wryly to his "winking needles."

Since 1946 Thomas had been plagued by debts and, above all, a devastating inability to work. Between 1946 and 1952, he produced only six poems and commented at the time, "It seems as if my faculties for self-criticism have grown more than my talent." Increasingly Thomas turned to alcohol. He suffered from gout, gastritis, shortness of breath, and frequent bouts of flu and bronchitis. In the early '50s his father was dying, his wife Caitlin was again pregnant, his children were ill, and his debts were huge. He needed money but could not stand to undertake the means to get it. His outward life became as turbulent and chaotic as his inward life: John Davenport described him as in "a state of terror."

In a letter to Oscar Williams, Caitlin remarked that Thomas had left off writing in favor of acting. She referred to Thomas's frequent trips to America, where he read his work—sometimes in a very drunken state—made some records, vowed he would return to Wales with riches, and played a part that he himself feared he could not sustain. Perhaps he wanted to escape from himself and, as Caitlin later said, "from the worst side of life. [He] deliberately wanted to do away with himself before he was forty."

Severe bouts of drinking preceded that of November 4. One time Liz Reitell found Thomas at New York's Algonquin Hotel temporarily deranged and ranting about war. Another time he drank whiskey and beer till the morning hours, then took Benzedrine. On November 3 he kept two drinking appointments, returned to his room, then suddenly went out again at 2 A.M. The owner of the White Horse later reported after a study of his inventory that Thomas could have drunk only six, at most eight, shots of whiskey. Thomas managed to get up on November 4, an unlikely feat had he really consumed 18 drinks, and went out to the White Horse again around midday for a few beers. On returning to his room he became wretchedly ill with gastritis and vomiting. Reitell attended him and called Dr. Feltenstein. He first gave Thomas an injection of ACTH, and the patient began seeing geometrical abstractions—circles, squares, and triangles. His situation worsened. Thomas's friend John Malcolm Brinnin reported, "As Dylan, raving now, begged to be put out, the doctor gave him a sedative." A half grain of morphine sulphate was given, hardly standard treatment for gastritis. If Thomas was suffering the breathlessness that he was prone to, the drug might have further cut off oxygen to the brain, which in turn induced the coma. At the time no one pointed the finger at Feltenstein. The doctor would not answer questions on the subject, and in 1974 he himself died, leaving the mystery unsolved.

?

How did the Pilgrims learn to grow corn?

Corn holds a vital spot in the birth of our country as the staple on which the first Pilgrims survived. With no farm animals and little aptitude for tracking game, the Pilgrims might *all* have starved (as it was, many of them did) had it not been for corn. Yet before landing on these wild shores, they had never seen it. Slowly developed from a wild grass of southern Mexico, corn was brought north by various waves of pre-Columbians. The Native Americans could produce hundreds of bushels of food per acre, far more than the produce of European fields, and with no wasted sweat and toil, no compulsion for rigidly neat rows, barren of weeds. Rather than struggle to fell massive trees, for instance, the Indians simply cut away a strip of bark all around unwanted trees, causing their leaves to die, thereby clearing the canopy for sunlight to reach the crops. The Indians did not break their backs with arduous plowing, but scratched a line in the earth with a pointed stick and sowed the kernels of last year's crop. More ingenious than the scarecrow was the whistle hung on a nearby tree, which sounded as the wind blew through it, alarming birds and mice that raided the seeds and young plants.

All this the Pilgrims learned from the Native Americans. "Plant the kernels of corn," they instructed, "when the oak trees' leaves are as big as mice ears." They sowed four seeds close together in a hillock, about a yard apart, and in the middle they placed a small fish, which decomposed and fertilized the crops. After the planting, the family dog apparently had to hobble around for a month on three legs—one paw was tied up near its neck to prevent its digging up and making off with the fish.

Not only did the Pilgrims make porridge, corn pone, ashcake, hoecake, and succotash (corn and kidney beans and occasionally dog meat) from corn, but even a kind of milk on which many infants were raised. This milk consisted of corn and the juices of boiled hickory nuts and chestnuts. The Pilgrims also learned to make hominy, first removing the outer layer of the kernel by soaking the corn in a wood lye substance, then crushing it in a hulling machine.

Most of this corn was soft flour corn, with large white kernels, or what we now call cow corn. Not until 1779 did the settlers come across sweet corn growing along the Susquehanna River in western New York. Now warring against the people who had helped them, the colonists carried some seeds from the Iroquois's crop back east, but it wasn't grown in any quantity until the mid-19th century.

The Peruvian Indians grew sweet corn in prehistoric times, according to James Trager, and it was also enjoyed by the Hidatsa, Iroquois, Mandan, Omaha, Pawnee, and Ponca tribes of North America.

?

How did Hirohito come to renounce his divinity?

On Near Year's Day, 1946, the Emperor of Japan addressed his people, now battered by war, shamed by defeat, and fearful for the position of their divine leader: "We stand by the people and we wish always to share with them in their moments of joy and sorrow. The ties between us and our people have always stood upon mutual trust and affection. They do not depend upon mere legends and myths. They are not predicated on the false conception that the Emperor is divine and that the Japanese people are superior to other races, and fated to rule the world."

This was music to the ears of General Douglas MacArthur, the Allied chief in the Pacific. As Supreme Commander Allied Powers (SCAP), with his office directly controlling the Japanese government after World War II, MacArthur could have ordered Hirohito's issuance of the above imperial rescript, but he asserts in his *Reminiscences* that it occurred "without any suggestion or discussion with me." Reports from imperial circles say otherwise, that an explicit order was delivered recalling the Emperor even as he addressed his ancestors at the imperial mausoleums at Uneba and Momoyama. Whether or not there was a direct order from MacArthur, certainly there was pressure, both internal and external, on Hirohito, who proceeded to act in the interests of peace.

Two significant factors, however, completely eluded those in Washington, led by former Secretary of State Cordell Hull, who clamored for punishment of the supreme scapegoat and demythification of the imperial notion. For one, Hirohito at the age of 14 had himself expressed doubts about his divinity, gently suggesting to his teachers what was anathema to the very life and sustenance of the country: that the legendary descent of the imperial line from god was biologically impossible. "I have never considered myself a god," he claimed in 1946. "Nor have I ever attempted to arrogate to myself the powers of a divine being." This was hardly the Emperor whom the West perceived, the Emperor who induced unflinching courage and binding loyalty, the divine being who, the Japanese military led us to believe, had encouraged the kamikaze pilots and fostered the belief that no greater happiness existed than to die in his service.

The steadfast reverence of the Japanese for their supreme leader was the other factor that westerners simply could not fathom. Some American authorities imagined (rather superficially) that once Hirohito, the man, with all his nervousness and human foibles, walked among the people they would see the truth, a Santa Claus out of costume, the end of a myth. This was not to be. Although some confusion and upheaval was inevitable, to this day a large percentage of the Japanese population tenaciously adheres to the tradition, honoring the Emperor as a divine being and peerless aristocrat.

The missing link between the Emperor's own apostasy and the faith upheld by the Japanese was the military, who intermittently from the 1920s campaigned to aggrandize his powers, to turn the Emperor into a shadowy symbol, remote, revered, inaccessible. The military could then glean the rewards of this loyalty as well as the power to speak directly where the Emperor, a puppet however divine, could not. The very idea of a divine leader in fact originated with the powers behind the throne who, with the abolition of the shogunate in 1868 and of the feudal estates, saw the need for a substitute authority and object of veneration to lead the people. They established a political and spiritual sovereign who, according to the Constitution of 1889, was "sacred and inviolable." All schoolchildren were taught that Japan "shall be reigned over and governed by a line of Emperors unbroken from ages eternal." They construed as fact the fantastical ancient legend (whose obscenities were deleted for the schools) of the imperial origin.

As the story goes, the god Izanagi and goddess Izanami were sitting up in the clouds in the dim, distant past. Some droplets of water gathered on the god's spear and fell to earth, where they congealed, forming an island. Izanagi and Izanami went down to explore the island, embraced, and gave birth to a son and a number of islands. These, of course, were Japan, the rest of the world being formed only later from flotsam and jetsam. Eventually, Izanami died giving birth to fire, but Izanagi heroically gave birth himself to three divinities: Amaterasu, the sun goddess, from his left eye; Tsuki-no-kami, the moon god, from his right eye; and Susa-no-o, god of the sword, from his nostril. This last was a rapacious god with a lusty appetite who violated even his sister, the sun goddess. Amaterasu in shame retreated to a dungeon, leaving the world in darkness until some warriors craftily aroused her jealousy with the pretense of another woman. Who was she? The warriors presented a mirror to Amaterasu, who unwittingly reappeared to challenge her beautiful rival. She gave birth to a son and ultimately her descendants came to earth to rule. The legendary first Emperor of Japan (600 B.C.) was—although his mother was a crocodile—the great-great-great-great-grandson of Amaterasu.

Thus, thousands of years later, the Emperor was still to be honored and obeyed. In Hirohito's time, every Japanese learned the Imperial Rescript on Education by heart, and bowed low when passing the Imperial Palace. When the Emperor appeared in public all heads were bowed and windows shuttered lest anyone glimpse his divine being.

According to Leonard Mosley, Hirohito came to resent the militarists who manipulated him and the fanatical field-grade officers who filled the people's heads with notions of ruling the world. In 1944 the Office of Strategic Services published a document that described the use of the Emperor's name as a "deliberate policy" to "reassure the people of the sacred nature of the war, and that the Emperor's divine destiny will assure victory." Furthermore, "the increasing use of the Emperor's name tends to make him share with the military the responsibility for the war."

No wonder that the Japanese, on the one hand, persisted in their faith and the Americans, on the other, saw the need to denounce it. Thus when an Imperial Conference decided to accede to the Potsdam Proclamation of the Allies to end the war, their acceptance carried a provision that the declaration "not comprise any demand which prejudices the prerogatives of His Majesty as a Sovereign Ruler." The camp was split in Washington between those who opposed the monarchy and those who felt that insistence on dissolving it would inspire fanatical resistance and prolong the war. Finally, President Truman issued the Allied reply, stating that "the authority of the Emperor . . . shall be subject to the Supreme Commander of the Allied Powers. . . ." This indeed upset the Cabinet, which feared for the national polity, and the Army and Navy Chiefs of Staff, who still wanted to fight to the end. Hirohito, resolute in his decision for peace, was forced to hold another Imperial Conference, where he repeated his acceptance of the terms. On August 15, 1945, the Japanese heard for the first time the voice of their Emperor, speaking haltingly over the radio to announce the end of war.

Thereafter the SCAP administration was established and slow, indirect pressure brought to bear on Hirohito. Most of his property was taken, his ministers tried as war criminals, his country shaken by riots. Just as MacArthur wished, Hirohito eventually came to him: ". . . to offer myself," said Hirohito with dignity, "to the judgement of the powers you represent as the one to bear sole responsibility for every political and military decision made and action taken by my people in the conduct of the war." It was just over three months later that Hirohito publicly renounced his divinity, and MacArthur was pleased.

?

How did they find a river in the Sahara?

For thousands of years stories circulated about a vast river without water concealed beneath the barren sands and dunes of the Sahara. Here, where rain falls only once in 30 to 50 years, the possibility of a riverbed seems ludicrous. But in November 1981 the space shuttle *Columbia* had a more penetrating look at the arid region than any who have traversed it by camel or on foot. The *Columbia* took a 50-kilometer-wide radar scan of the desert, and because there was no moisture to disperse the waves, they passed beneath the surface to a depth of 5 meters, finally reflecting from the bedrock. These signals were processed by a computer and analyzed by scientists from the U.S. Geological Survey, the Egyptian Geological Survey and Mining Authority, and a number of other institutions. Released in December of the following year, the images showed an astounding network of waterways, floodplains, and former expansive river valleys throughout southern Egypt and northern Sudan.

Geologists believe the Sahara dried out during an ice age 2 million years ago. It did not, however, remain in this state (similar to the one that we know today) throughout these many years. The region enjoyed intermittent rainy periods, 200,000, 60,000, and 10,000 years ago. The presence of water drew animals and men, who found life within it and along its shores. Guided by the revealing images, a United States–Egyptian team dug along the banks of an ancient river in September 1982 and confirmed what the pictures had told. Here lay tools and artifacts reportedly used by *Homo erectus*, an early man, who lived and hunted in the lush Sahara of 200,000 years ago.

?

How did they make Warren Harding the Republican presidential nominee in 1920?

With his genial disposition, lack of political drive, and tastes for wine, women, and poker, Warren Gamaliel Harding was not only a startling choice in 1920 but one of the least qualified candidates ever nominated by the Republican or Democratic party. But the young politician from small-town America—Blooming Grove, Ohio—had good looks, a pleasing voice, and a lot of charm. He did not antagonize. He played a game of compromise, made few enemies, and let his opinions drift and change with the tides of public opinion. He supported women's suffrage, although he didn't believe in it. He opposed excess-profits taxes and supported high tariffs to stay in the good graces of the Old Guard.

Warren Harding keeps a pressing schedule as President. Here he (right) *and Harvey Firestone scan the news while Thomas A. Edison takes a nap.*

Although not unduly swept up in the Red Scare that racked the country, he was suspicious of labor unions, and in the spirit of the Sacco and Vanzetti affair, he supported restricted immigration and thought anarchists should be shot.

Originally, Harding played his unobstrusive political game not to become President, but because he liked being a senator. Although his attendance was pitiful—he missed 43 percent of the roll calls from 1915 to 1919—he enjoyed the prestige, entertained weekly in his Washington home, and one way or another made friends with influential people. The presidency was, in fact, so far beyond the scope of his desires that when in December 1919 he announced he would run, he commented to his friend F. E. Scobey: "The only thing I worry about is that I might be nominated and elected. That's an awful thing to contemplate."

The overriding mood of the Republican convention that began in Chicago on June 8, 1920, during a stultifying heat wave was confusion. And this proved fertile ground for a legend to arise and persist for years to come—a legend of boss control, financial manipulation, and a "Senate cabal" plotting through the night of June 11 in a "smoke-filled room" at the Blackstone Hotel to force the Harding nomination. For the first day of balloting, June 11, had seen a deadlock between the two top contenders, General Leonard Wood, standing for nationalism and militarism, and Governor Frank O. Lowden of Illinois: in the fourth ballot, Wood had 314½, Lowden 289, Harding a meager 61½. The tremendous turnabout the following day could only be explained as the work of a senatorial clique that decided to make Harding its puppet—or so wrote George Harvey, publisher of *Harvey's Weekly*, an editor and publicity maker who was on the scene, sharing the legendary suite of Republican National Committee Chairman Will Hays at the Blackstone. Harvey maintained that Harding was called to the room at 2 A.M. and asked if there was any reason that he should not accept the nomination. When Harding replied there wasn't, the "boys," eager for a president they could control, proceeded to get him the requisite votes.

It's a catchy story, but far from the truth. If not the bosses, who was behind Harding's precipitous rise? For years there were others, friends and opportunists alike, who worked behind the scenes to assure Harding's victory, first in Ohio, then in Chicago.

High on this list, if not at the top, was Harry Micajah Daugherty, a shrewdly aggressive lawyer, keen political strategist, and influential Republican. The two men met on the way to a privy in the backyard of a small Ohio hotel around the turn of the century. Daugherty later claimed to have thought then and there, "What a President he'd make!" When the lawyer's own political career was cut short in 1916 and when Teddy Roosevelt, a likely Republican candidate, died in 1919, Daugherty decided to make a president of Harding—both for himself and for the regular faction of the Republican party. "I found him," boasted Daugherty, "sunning himself, like a turtle on a log, and I pushed him into the water."

Other early supporters were F. E. Scobey, former Clerk of the Ohio Senate, Charles Hard, Secretary of Ohio's pro-Harding Republican State Advisory Committee and owner of the *Portsmouth Daily Blade*, and George B. Christian, Jr., Harding's private secretary. They tackled the first major problem: getting Harding to run. Initially opposed to the idea, Harding began to entertain it by the fall of 1919, although his wife, Florence, thought it a terrible mistake and fiercely fought off the Senator's persistent friends. Then in October 1919, pressure in the political arena within Ohio precipitated Harding's decision. Threat of a coalition between the Cincinnati and progressive factions of the Republican party and increased support for General Wood led antiprogressives—such as Harding—to fear for their political survival in Ohio itself. Feeling he had no choice, the Senator announced his decision to run for President. The campaign was on.

Daugherty ran himself ragged with letters and phone calls. Harding corresponded with newspapermen and his influential friends. In Boston he delivered his famous "Back to Normalcy" speech calling for "tranquillity at home," "thoughtful labor," "sober capital," and "wholesome common sense." "Gamaliel represents . . . capitalism, openly and unashamed," wrote H. L. Mencken. "He is . . . a frank reactionary. Well, if we are to have reaction . . . why not hand over the conduct of the state to the honest reactionary?"

Harding and his supporters scraped together their pennies for his modest campaign, relying entirely on contributions. Wood, on the other hand, had millions at his disposal and was not shy about using them. His outlay of cash eventually aroused such bitterness that a Committee on Privileges and Elections (later called the Kenyon Committee after its chairman, Senator William S. Kenyon) was established to investigate. Its findings were a boon to Harding, who, with the lowest campaign expenses (only $113,109), appeared clean. Wood, having spent a staggering $1,773,303, was criticized for trying to buy his way in, and the Lowden contingent was charged with bribing the delegates. These findings would prove significant on that fateful Saturday, June 12, at the convention.

To the end, Harding maintained a low profile. By so doing he managed to obtain considerable hidden support—that is, where he did not show up as the first choice, he figured strongly for the second or third. When the convention opened, over 40 minutes of cheers and stomping followed the presentations of Wood and Lowden. Harding got only ten minutes' worth, but Frank Willis, who nominated him, left an impression of friendliness and good humor—a welcome relief to the heated and confused convention.

Although it was not a secret meeting, there was activity in Hays's suite at the Blackstone that night. Disorganized and indecisive senators streamed in and out between 8 P.M. and 2 A.M., wondering what was to be done. In addition to the Wood-Lowden deadlock, problems could be found with all the other candidates (who included Herbert Hoover, Calvin Coolidge, Hiram Johnson, and Will Hays himself); they were too young, or too progressive, or too bland, and so on. Harding loomed up as a compromise. The following morning, pamphleteer George Harvey and Senators Reed Smoot and Charles Curtis apparently commented to the press that Harding's selection seemed likely. Thus the public would hear of a "Senate cabal" and its decision. But on the first ballot Saturday morning, 13 of the 16 senatorial delegates at the convention still voted against him.

With each ensuing ballot, though, Harding gained ground. In many instances, the second-choice support began to show itself, and the staunch loyalty of John B. Galvin, one of the Big Four Ohio delegates, was a boost in Harding's favor. During a midday recess Wood and Lowden rode through the Chicago streets in a taxi, each trying frantically to get the other to step down and take the vice-presidential nomination—but without success. By the afternoon session, many of the baffled delegates were ready for a compromise, a safe bet, a winning smile. Connecticut voted overwhelmingly for Harding, followed by Kansas and a decisive state, Kentucky. Pandemonium erupted and other states turned in favor of Harding. He was well ahead in the ninth ballot and tumult sufficed for ratification on the tenth. "Rather than boss control," writes Harding's biographer Robert K. Murray, "*no* control was the final key to Harding's nomination success."

What did the nominee himself have to say about it? "We drew to a pair of deuces," he remarked, "and filled."

?

How did they design the Model T?

As simply as possible, for the most economical and efficient production to date. This was the car for the masses, the car that revolutionized the industry in its early days, that ten years after its appearance sold for an average of $400 retail while Ford netted over $80 million annually. "You can have any color so long as it's black," quipped Henry Ford, who for years before the car's birth in 1908 had dreamed of a "universal car," which would eliminate the frills and high production costs that previously had confined cars to the garages of the super-rich. For several years his closest colleagues, including Charles Sorensen, foreman of the pattern shop, and Joe Galamb, head of transmission engineering and design, brainstormed in secrecy in a special room set aside for this purpose. The beloved Tin Lizzie that resulted was made possible in part by the introduction to the United States of English vanadium steel at that time. This light steel, whose tensile strength was triple that of the steel then used, was first produced in this country at the United Steel Company's plant in Canton, Ohio, in 1905, and Ford was quick to perceive its possibilities. With it he could, and did, produce a lighter, cheaper, and better car.

Perhaps more significant for the history of industry and for consumers of the early 20th century than the materials used was the novel manner in which they were assembled. Prior to the

Unprecedented speed, accuracy, and economy were the trademarks of Ford's Model T assembly lines at the Highland Park plant in 1914.

Ford's 1909 Model T was homely, useful—and revolutionary. In the decades to follow, millions of families became proud owners of the rugged Tin Lizzie.

Model T, each car was built by hand. The chassis was stationary and each part designed for that car was brought to it by skilled workers. Ford realized that time and money could be saved by incorporating and improving upon mass-production techniques used sporadically in the past. Eli Whitney's machines had rapidly produced large numbers of interchangeable parts for muskets as early as 1798. The Chicago meat-packers of the 1860s used conveyor belts that transported the carcass from worker to worker, each of whom performed one task. Elihu Root furthered the idea of dividing and simplifying steps in the manufacturing process, thereby achieving faster production of Colt six-shooters in 1849.

Ford, for his part, got his cars into stands that moved from one team of workers to the next, but these cars were still hand built. So he developed machines to make masses of interchangeable parts at high speeds and then machines to assemble the parts, finally achieving perfect synchronization between the moving assembly line with its growing car and the surrounding feeder lines. This could first be seen in 1913 at his famous plant at Highland Park, Michigan, covering 60 acres, containing 15,000 machine tools—presses to stamp out steel frames, elaborate systems of chutes, gravity slides, and conveyors. These were attended by unskilled laborers who by the early '20s earned a minimum daily wage of $5.00—double what they took home before the successes of the Model T. In its first year the Model T sold over 10,000 units; over 2 million were bought each year in 1923, 1924, and 1925; and Ford had sold a total of over 15 million by 1927 when the car was discontinued.

At top speed the Tin Lizzie sailed along at 45 miles per hour, getting 20 miles to the gallon. It had a 20-horsepower, 4-cylinder engine, weighed 1,200 pounds, and cleared those rustic early roads by 10½ inches. Wherever possible, the light, strong vanadium steel was used: in the crankshafts, connecting rods, pistons, pins, and valves, for example. Ford and his colleagues strove for the simplest engine design possible; the crankcase and cylinder barrels were in one unit, and for the first time, a gas engine was designed with a separate, detachable cylinder head. The car had three foot pedals: one for the brake, two for the two-speed transmission. Ford incorporated only the most efficient elements of his former models, streamlining and developing new units for his assembly-line production. The result: Americans began for the first time to view cars as feasible, and then necessary, not to mention desirable, possessions—yet another inalienable right.

?

How did they discover Troy?

Heinrich Schliemann, a self-made millionaire with a lifelong belief in the historical accuracy of the Greek classics, discovered ancient Troy, the site of the battles fought in Homer's *Iliad*. Born in 1822 in Mecklenburg, Germany, Schliemann was introduced to the *Iliad* and the *Odyssey* by his father, a Protestant minister. As a boy Heinrich would listen to the story of the siege of Troy, brought about by the Trojan prince Paris's abduction of Helen, the wife of King Menelaus of Sparta. It is common knowledge that the Homeric epics are based on myth and legend, having been composed, perhaps, by a number of poets over a span of time, not by a single individual. To Schliemann, however, the poems' description of the Trojan War was historical fact, and the young Heinrich dreamed of discovering Troy.

The dream was slow to materialize. At the age of 14 Schliemann was apprenticed to a grocer, which prevented completion of his formal education. He lost this position at the age of 19 as a result of illness and boarded a ship bound for Venezuela. The ship was wrecked in the North Sea, but Schliemann survived. He became a clerk in Amsterdam and began to educate himself by learning a number of languages, including English, French, and Dutch. His knowledge allowed him to get a job with an Amsterdam-based company of indigo merchants that traded with Russia. After being sent to Russia by his employers, Schliemann set himself up in 1846 in St. Petersburg as a wholesale indigo merchant and began to make his fortune. In 1850 he went to California during the gold rush and formed a banking agency in Sacramento, increasing his wealth. Two years later he returned to St. Petersburg, married, and profited from dealing in war matériel during the Crimean War. He then acquired a knowledge of Latin and Arabic during a tour of the Mideast, and later learned modern and classical Greek. "In the midst of the bustle of business," he said, "I never forgot Troy. . . . I loved money indeed, but solely as a means of realizing this real idea of my life." In 1863 Schliemann retired from business and settled in Paris to study archaeology. Five years later he set out for a tour of ancient Greece and began his search for Homer's Troy, with its "wide streets" and "lofty towers."

Other explorers had looked for Troy at Bunarbashi, a village situated on a plain in northeast Turkey, but their excavations had yielded nothing. Schliemann visited Bunarbashi, but did not find one cold and one hot spring, which were said by Homer to mark the city of Troy. Instead, he found 40 cold springs. Schliemann traveled around the plain until he spotted a hill called Hissarlik, located three miles to the south of the Dardanelles. This location would make it easy for the Greeks to travel back and forth from their coastal camp to Troy several times a day, as described in the *Iliad,* whereas Bunarbashi was eight miles from the coast, making swift forays difficult. There were no springs at Hissarlik, but these, Schliemann believed, could have dried up over time. Also, the circumference of the mound at Hissarlik was small enough for Achilles to have chased Hector around it three times, as the *Iliad* said.

Believing that he had found Troy, Schliemann left Turkey to take care of personal matters. (His marriage was failing, and he obtained a divorce in Indianapolis.) Returning to Hissarlik in 1870, Schliemann dug a trial trench into a corner of the mound and found a massive stone wall 16 feet below the surface. He applied to the Turkish government for a permit to excavate, and while he waited for the permit he remarried, this time to Sophia Engastromenos, a 17-year-old Greek woman who would later take part in all aspects of Schliemann's excavations. The permit arrived in September of 1871, and on October 11 the digging began. Schliemann hired 80 laborers to cut a trench 33 feet into the mound, tearing through any obstacles, including walls and foundations, that they encountered along the way. The digging was halted after six weeks because of the weather, and was resumed the following March.

During the next season a labor force of over 100 worked to uncover layers of the mound; they hit bedrock at a depth of 45 feet. Schliemann assumed that King Priam's city lay at the base of the mound, and by 1873 he had recognized seven cities of Troy, built one on top of the other. He found that the earliest city was too small and primitive to be the Troy of the *Iliad,* and he concluded (wrongly) that the third city from the base was Homer's Troy. This level contained burned masonry and scattered gold, silver, and copper artifacts. Schliemann decided that this was the city that had been destroyed by the Achaeans at the end of the Trojan war.

One morning Schliemann spotted gold in loose earth not yet examined by any of his workers. He sent all the workers home that day and dug out by hand a number of gold pendants and earrings, pieces of ivory, and chains and brooches. Convinced he had found the treasure of the king of Troy, Schliemann said, "Apparently someone in Priam's family had hastily packed away the treasure in boxes. Then, on the walls, this person met his death." Schliemann, like everyone else at the time, was not aware of the actual dates of the Homeric epics. He was looking, simply, for the city that was, according to the *Iliad,* sacked by the armies led by King Agamemnon. Today it is believed that the treasure Schliemann found actually dated from the 22nd century B.C., not from the 8th century B.C., when the *Iliad* and *Odyssey* were given their final form, or from the 13th century B.C., when the events in the epics probably took place.

According to the terms of his permit, Schliemann was required to give half of his findings to the Turkish government. Instead, he smuggled the gold artifacts out of Turkey and into Greece. Then he wrote an article for a German newspaper in which he admitted to the smuggling. The Turkish government sued, and Schliemann ended up paying a fine and donation totaling $10,000 in 1875.

In 1876 Schliemann turned his attention from Hissarlik to Mycenae, at the north corner of the plain of Argos in Greece, which he believed was the burial ground of Agamemnon. At Mycenae, Schliemann uncovered five graves containing the largest archaeological treasure found before the discovery of Tutankhamen's tomb in 1921. What Schliemann did not know at the time was that he had found the remains of a Mycenaean civilization that had thrived during the 16th century B.C., and that Agamemnon, if he indeed existed, would have lived perhaps at the end of the 13th century B.C.

In 1878 Schliemann resumed excavations at Hissarlik, aided by the German scholar Rudolf Virchow. In addition to giving expert advice, Virchow persuaded Schliemann to give the gold artifacts, then stored in Greece, to Berlin. Unfortunately most of this treasure vanished during the Second World War. In later seasons, from 1882 to 1890, the architect Wilhelm Dörpfeld joined Virchow and Schliemann at the Troy excavation. By 1890 two more levels had been added to the seven initially identified by Schliemann. Today it is thought that cities I to VII existed during the Bronze Age, around 3000 B.C., city VIII in the Geometric period (700 B.C.), and city IX (Ilium) in Hellenistic and Roman times. And it is believed that the Trojan War was but a chapter in Mycenaean expansion, part of an effort to gain control over an area of strategic importance.

Schliemann never realized his last project. He planned to excavate the palace of King Minos at Knossos on the island of Crete, where Theseus, according to legend, traveled to slay the Minotaur. But before Schliemann could obtain a permit to dig on Crete he died suddenly in Italy, in 1890, at the age of 68. He was by then an international celebrity.

The excavations at Troy were carried on by Dörpfeld until 1894, then resumed from 1932 to 1938 by a group from the University of Cincinnati led by Carl William Blegen. Dörpfeld found that the sixth city, not the third city, was the largest and best built, and he thought it had been destroyed by an invading army. Blegen, however, found that the sixth city had been leveled by an earthquake, and that the seventh city had been destroyed by fire. The fire, Blegen discovered, was accompanied by violence, as evidenced by the damaged bones at the site. Also, the inhabitants of the seventh city had stored away much food, as if preparing for a siege. Blegen concluded that the seventh city was the one described in the *Iliad* as King Priam's Troy, the one sacked by the Achaeans, and his conclusion is accepted as accurate today.

?

How did Superman fly?

When Superman speeds across the heavens to save Lois Lane or fearlessly dives and swoops around the skyscrapers of Gotham or over the San Andreas Fault, kids really believe he's flying and many of the rest of us come close to believing, too. That's because the moviemakers put a lot of effort into not only getting Christopher Reeve airborne, but making the action of flying and all it entails really convincing, right down to the rippling of wind through his broad red cape.

Getting actors and actresses off the ground by means of camouflaged wires is nothing new. But whereas in the past the flyer was suspended on only one wire and had little maneuvering ability as a result, a new system involves two wires, one from each hip. Superman wore a fiber-glass hip harness—or plastic underwear lined with fleece—and to this the wires were attached via ball-bearing swivel joints. High overhead the wires met a bar with a supporting system of cables and sheaves, operated by hydraulic rams. There, too, was a track along which the wires were moved, conveying Superman around the studio—with some near misses of the sets—landing our hero in a net. The wires were painted blue when Superman flew against a blue screen, and sometimes a completely new background was substituted so that no wires would be seen. A crane specially built for the effect was 250 feet high, and Reeve courageously veered around the studio, often at heights of 200 feet.

Despite Reeve's exceptional skill at this—something to consider when it is usually inanimate objects we see hurtling through the heavens in films these days—there was initially something not quite right about his flying. The cape wouldn't swirl properly and no wind machine was capable of making it do so. Les Bowie, responsible for matte painting in the film, came up with an unprecedented solution. He made a device, controlled by radio, that operated long rods and these in turn were attached by lines to Superman's cape. Somehow the various motions of the rods could make the cape billow and furl as naturally as clothes hung out to dry on a windy day.

Not only did Christopher Reeve have to fly well, he had to cover distance against a realistic background. This feat was accomplished by a special-effects team which, besides Bowie, included creative director Colin Chilvers, Roy Field, responsible for opticals, Derek Meddings, Denys Coop, and Geoffrey Unsworth. Sometimes the action shots of Reeve in flight were combined with pictures of models of various backgrounds or still shots of actual scenes. Miniatures and models are used profusely in films, not only for landscapes and cityscapes—Rome in *Ben Hur,* for example—but also for ships and submarines, for battles and wrecks where full-scale casting would be impractical, and for spaceships—yes, even those in *Star Wars* were models, drawing on over 300 model kits and costing a mere $8 million to produce. The trick with models and miniatures is to provide meticulous detail and also adequate size to help surmount the impression of reduced scale. Sharply contrasted light and shadow can also help make a small object appear huge on the screen.

One technique of combining the action of Superman in flight with a still of a city (let's call it Gotham) is known in the trade as a traveling matte. Superman is photographed in action against a blue background. An exposure is then made which, in effect, is a silhouette. Called a matte, this opaque mask is "bipacked" with the image of Gotham—that is, it is placed in contact with the background scene using a process camera that can hold two rolls of film simultaneously. At first, the matte appears on the background scene as a blank "hole" in the shape of Superman; it moves across the background scene changing shape and position in each succeeding frame. At last the live action of Superman is printed into the space occupied by the matte with the aid of an optical printer. This machine comprises a movie projector and a process camera that face each other; the former projects positive film, the latter negative. Although its primary use is for making duplicates, it can do all kinds of tricks with multiple images, reverse action, skip frames, and zooms. In this case, it sets Superman aloft, propelling him at speeds clearly evident in the passing cityscape and his wind-lashed cape.

?

How did they first measure the weight of the earth's atmosphere?

What would the weather forecast be without a map of highs and lows? How could the weatherman guarantee us a sunny weekend or predict a blizzard if it weren't for the barometer? This instrument is undoubtedly one of the meteorologist's most useful tools, but there's nothing magical or complex about it—its precursor was invented over 300 years ago.

In 1644 a protégé of Galileo's, Evangelista Torricelli, filled a glass tube with mercury and submerged one end in a basin of mercury, while keeping his finger over the other end. When he removed his finger, some of the mercury remained suspended in the tube rather than sinking to the bottom and slipping into the basin. The reason, according to Torricelli, was that "on the surface of the liquid in the basin presses a height of 50 miles of air." Atmospheric pressure was, in fact, keeping the mercury suspended in the tube and Torricelli watched its level change with different weather conditions.

"We live," remarked the inventor in a reflective mood, "submerged at the bottom of an ocean of elementary air."

?

How did Neanderthal man find food and shelter?

Paleontologists believe Neanderthal man to have lived in the world from 100,000 to 40,000 years ago, a period that began in a warm interglacial phase but included the last ice age (70,000 years ago) as well. Thus these primitive humans, found predominantly in Western Europe and the Mideast but as far north as southern

Siberia, confronted a wide range of climatic conditions and survived acute cold in some very inhospitable regions. To do this they must have developed a shrewd knowledge of the natural world, as well as the quite sophisticated technology that is manifested in their artifacts.

The sensational images of *Quest for Fire* aside, our primitive forebears had fire, and they built shelters with proper ventilation to accommodate indoor hearths. Wood provided fuel and building material for the shelters, and where this was scarce, animal bones were substituted. In many instances the Neanderthals lived in caves. The size of their communities has not been conclusively shown, but the number was likely determined by how many the local resources could support, and may have averaged about 25.

The Neanderthals left their predecessor *Homo erectus* far behind when it came to toolmaking. Their culture, broadly characterized as Mousterian after Le Moustier, a cave in the Dordogne, France, had tool kits with some 60 identifiable items. Among them were knives, scrapers, and projectiles, fashioned by the Levalloisian technique of flaking. A single rock was split into a number of flakes, each of which became a core from which a tool was shaped. *Homo erectus,* on the other hand, would have obtained only one tool per rock, using the Acheulean technique of simply paring down the stone to the desired shape. According to Richard E. Leakey, the Levalloisian technique could produce seven feet of cutting edge from 2.2 pounds of flint—five times as much as might be generated by the Acheulean method.

Some of these tools were useful when it came to hunting, although the Neanderthal favored gentle herbivorous mammals whenever possible. Goat, sheep, wild cattle, pig, and land tortoise were common game; bear, deer, fox, and marten posed a bit more of a challenge. A group of hunters would run a herd over a cliff, or into dead-end areas such as canyons where the animals could be caught and slaughtered. Single-handed hunting may have occurred, but the communal approach is more likely, according to Professor Ralph S. Solecki of Columbia University, who excavated Neanderthal bones in the Shanidar Cave in Iraq.

Solecki's findings, together with those of Arlette Leroi-Gourhan of the Musée de l'Homme in Paris, add a touching note to the culture and sensibility of the primitive men, once thought to be little more than apes. We find here the first instance of ritual burial. Among the food and tools buried with a person, to sustain him in death or in the passage to another life, were flowers. Whether for their medicinal properties or simply their beauty and comfort, yarrow, cornflower, grape hyacinth, hollyhock, and woody horsetail were collected and spread like a blanket on which to lay the dead. The novel image of these people gathering at the graveside with bunches of blue and yellow flowers suggests a spiritual and emotional world at odds with our popular conceptions of the Neanderthals.

?

How did 3M tape become Scotch?

In 1925, when Detroit was alive and well with a burgeoning auto industry, some car manufacturers were producing two-tone models. Flashy, yes, but a nightmare to paint. So the car makers turned to the Minnesota Mining and Manufacturing Company (3M) for a sturdy tape they could run along a seam while painting to keep clean the border where the colors met. The 3M people bought the idea, but because the cost of the adhesive was high, someone skimped—the tape went out with a strip of adhesive missing from the center line. Naturally, it didn't stick and the paint dribbled into all the wrong places. Irate auto workers attacked the 3M salesmen: "Take this tape back to those Scotch bosses of yours and tell them to put adhesive all over the tape. . . ." The bosses made a better tape, but the name Scotch stuck.

?

How did they discover the Hope diamond?

No other gem in history has provoked such fascination and mystery as the deep-blue 45.52-carat Hope diamond. This stone, however rare and beautiful, allegedly brought evil and misfortune to all who wore it. That, legend claims, is why Louis XIV died in disgrace; Louis XVI and Marie Antoinette fell beneath the blade of the guillotine; the beautiful Princesse de Lamballe was torn to pieces by a mob of Frenchmen; a Folies-Bergère actress was shot onstage by Prince Ivan Kanitovsky, who himself had given her the stone. Even the origin of the diamond is embroiled in legend, and one that links the diamond to the divine. A French merchant, Jean Baptiste Tavernier, allegedly tore the stone from an idol's eye in India—and was later ripped apart by wild dogs on the steppes of Russia.

These are among the myriad myths and events that are unsubstantiated or have been disproved by the Smithsonian Institution, where the inauspicious stone now resides. According to the Smithsonian, the known history of the Hope diamond dates only to the early 19th century. A London diamond merchant, Daniel Eliason, sold the diamond to Henry Philip Hope for £18,000 (about $90,000), a considerable sum back in 1830. Hope came from a long line of London bankers and, having sold Hope and Company to the Baring firm in 1813, had plenty of spare cash to indulge his fancy for art and precious stones. His extraordinary new diamond had apparently been in England for some years, for a corroborating description and illustration are found in John Mawe's *A Treatise on Diamonds and Precious Stones*, published in 1823, and reference is made to a "superlatively fine blue diamond of above 44 carats" in Mawe's first edition, published in 1813. Furthermore, American gemologist George Frederick Kunz found in Quaritch's bookshop in London a book by Pouget containing two precise drawings of the Hope diamond. These were made by a Soho lapidary in 1812.

Twenty years before this, the world possessed another spectacular diamond, also the color of sapphire. This was the French blue, part of the French crown jewels. The merchant Tavernier had traveled to India in 1642 and bought, among others, a diamond weighing 112 3/16 carats, believed to have come from the Kollur Mine in Golconda. Louis XIV was intrigued with Tavernier and still more intrigued with his wares. For the blue diamond alone he paid 220,000 livres. Because the Indian cut was rather rough and failed to highlight the stone's brilliance, the King had his diamond cutter, Sieur Pitau, recut it. The finished gem, 67 1/8 carats, was triangular and was named the Blue Diamond of the Crown. The Sun King wore the diamond suspended on a long blue ribbon around his neck; Louis XV had it set in the Order of the Golden Fleece; and Louis XVI wore it along with his diamond epaulets and jewel-studded sword. But so much grandeur and display was its own ruin. The Revolution erupted in France and the King and his ostentatious Queen, Marie Antoinette, were whisked off to prison. The Minister of Interior then suggested to the National Assembly that the jewels be sold for money to back the inflationary paper currency. In September 1792 some thieves beat them to it, slipping into the Garde-Meuble (a temporary treasury) and making off with nearly 3 million francs in jewels. While most of the stones were soon found under a tree, the French blue was never recovered.

The Hope diamond's appearance two decades later got a London jeweler, Edwin W. Streeter, to thinking. If the French blue had been cut to disguise it, the result might well have included the smaller Hope diamond, one of whose sides is straighter than the other, perhaps marking the plane of cleavage. Streeter came across another clue in 1874. At the sale of the jewels of William, Duke of Brunswick, he noted a deep-blue, drop-shaped diamond of six carats. He compared this with the Hope and found their colors identical. As far as Streeter and many others were concerned, this proved beyond doubt that the French blue had been recut and had become the Hope. Just one of the flaws in this theory (which Susanne Steinem Patch points out in her book on the Hope) is that if the 6-carat and the 44-carat diamonds were *truly* of the same color, they would not *look* identical because of the difference in size, or depth, which would produce different intensities of color. It is possible Streeter looked only at the quality and called the two shades identical, despite the disparity in intensity. But we cannot know for certain on what he based his appraisal. Furthermore, Streeter discussed his theory about the Hope's origins in a half-dozen or more editions of his *Precious Stones and Gems* and in *The Great Diamonds of the World*, and some of his comments are blatantly contradictory. For example, stones described in one place as of a different color than the Hope are elsewhere described as part of the same diamond that yielded the Hope.

There are still missing threads in the elaborate tapestry of the Hope's history, and for the rational historian its origins are yet to be discovered. It seems that the prized stone, once bearing so much influence, refuses to relinquish all its mystique.

?

How did they decide to publish *Time* magazine?

Time magazine was the child of two ambitious Yalies who grew up in a time and of a class that cultivated self-assurance and initiative. The first issue hot off the press in March 1923 appeared when the two men behind it were only 24 and 25. They were Henry Robinson Luce, born in China on a Protestant mission, and Briton Hadden of Brooklyn, the son of a stockbroker, grandson of the president of the Brooklyn Savings Bank. Both boys, born in 1898, reached the Hotchkiss School in 1913 and Yale three years later. Both overcame stiff competition to work on the *Yale Daily News* and Hadden was readily chosen as chairman. When he graduated, the class correctly voted him "most likely to succeed" and "hardest worker," while Luce received "honors of the first rank."

Even before graduation, the seed was planted for a collaborative enterprise down the road. In 1918 the two sophomores were stationed in a desolate military post in South Carolina. Recalled Luce at a dinner on *Time*'s 20th anniversary:

> There's a picture in my mind . . . of an army camp in the last war; of two underaged second lieutenants, Brit Hadden and Harry Luce—two shavetails, two second looies doing training duty down in Camp Jackson, South Carolina. . . . One night Brit and I were walking back to our barracks through the vast, sprawling camp. At each step, our feet sank ankle-deep into the sand. But we ploughed on for hours—and talked and talked. . . . We were talking about "that paper"—about something we would do—cross our hearts—some day. . . . I think it was in that walk that *Time* began. . . . Why do I say that *Time* was born during that long walk, of which I cannot remember a single phrase or sentence? On that night there was formed an organization. Two boys decided to work together. Actually that had probably been decided long before then. But that night seemed to settle it.

On graduating from Yale, however, the two went their separate ways: Luce to Oxford, England, then to Chicago as a reporter for the *Daily News;* Hadden to New York to report for the *World*. Fate soon brought them together on the *Baltimore News,* where they worked for three months, tasted the publishing world, gleaned something of the business side, and took off for New York to launch their dream in February 1922.

Initially this dream was a paper called *Facts,* intended, as the young men's prospectus stated, "to serve the modern necessity of keeping people informed. . . ." The title was abandoned when Luce, tired after a hard day, was struck by a subway ad—it was either TIME TO RETIRE or TIME FOR A CHANGE—and Hadden readily took to the new idea. The unique angle that would set *Time* apart from the masses of other newsprint was Luce and Hadden's insistence that "people are uninformed BECAUSE NO PUBLICATION HAS ADAPTED ITSELF TO THE TIME WHICH BUSY MEN ARE ABLE TO SPEND ON SIMPLY KEEPING INFORMED."

As Robert Elson points out in his comprehensive history of Time Inc., South Carolina might seem an unlikely spot for this idea to germinate, yet the degree to which people there were uninformed must have made a marked impression on the young men. Two or three talks a week were arranged just to tell the soldiers what the war was about. At the time, only one child in seven finished high school and far fewer, of course, went on to college.

"*Time* is interested," wrote Hadden and Luce, "—not in how much it includes between its covers—but in HOW MUCH IT GETS OFF ITS PAGES INTO THE MINDS OF ITS READERS. . . .

"It differs from other weeklies in that it deals *briefly* with EVERY HAPPENING OF IMPORTANCE and presents these happenings as NEWS (fact) rather than as 'comment.'"

The first eight months in New York were arduous ones, devoted to writing the prospectus, raising money, learning about selling subscriptions, and devising a new writing style. Fortunately, friends proved generous with time and money: by October 1922, over $86,000 had been pledged; of the original 70 shareholders, 46 were Yale men, 14 were classmates, 14 belonged to the elitist club Skull and Bones. The surprise star behind the venture proved to be Mrs. William L. Harkness, mother of a Yale classmate, who came forth with $20,000 without blinking an eye. At the end of November a judge also by the name of Luce (a distant cousin) aided in incorporating the company, free of charge.

All was not rosy, though. The young men and their small staff worked steadily without a day off in their dreary office space, first on East 17th Street, then at 461 8th Avenue., and finally at 9 East 40th Street, where, despite the respectable location, they got no heat on weekends.

After two trial issues, the first issue of *Time*, "the Weekly Newsmagazine," was published. A slim magazine of 32 pages, including covers, it was designed to be read in an hour, and sold for the then-high price of 15 cents. The cover sported not a movie actress but 86-year-old Joseph G. Cannon, then retiring from Congress. Inside were 22 departments, including "National Affairs," "Foreign News," "Books," "The Theater," "Law," "Religion" (written by Luce), and others still featured today, as well as "Imaginary Interviews" with celebrities and commentaries called "Point with Pride," and "View with Alarm," which directed the reader to look, perhaps with a different viewpoint, at another page in the magazine. Editors Luce and Hadden had four staff writers and various contributors, including Archibald MacLeish.

Luce was pleased with the first issue, and hopes were high to meet the projected sales of 25,000—the entire printing. Actual sales fell far short, however: 9,000, with newsstand returns numbering 2,500 out of 5,000.

Needless to say, the editors persisted. (Hadden, however, died in 1929 at 31.) There were upheavals, fights, and fiascoes, but the original idea ultimately proved a winner. Today, *Time* magazine sells over 4 million copies weekly, bringing in annual advertising revenues of over $250 million, and the Time Inc. empire provides work for some 25,000 people

?

How did they find the poems of Emily Dickinson?

Emily Dickinson was born in 1830 in the quiet, respectable town of Amherst, Massachusetts, and died there 56 years later, her genius as yet unrecognized, her manuscripts, like her life, safely hidden from the world. As daughter of the treasurer of Amherst College, Emily might have taken part in society but chose not to, turning increasingly to a life of solitude. "I do not go from home," she wrote when only 23, embodying to an uncanny degree the self-sufficiency and introspection of New England transcendentalism. Unknown to her family, Emily wrote prodigiously in her room throughout the long, cold winters, showing—whether from fear, or vanity, or miserly protection of her art—only light pieces, some nature poems, and elegies to a few friends. Her wit and penetration displayed themselves in conversation (though her austere father frequently disapproved), and in scattered epigrams and enigmatic phrases written to friends, often with requests to burn these letters before foreign eyes could find them.

Probably the closest person to Emily Dickinson was her sister, Lavinia, a brilliant conversationalist with a passion for flowers, who became an eccentric spinster alongside the reclusive Emily and saw to many of her practical needs. So in 1886, when Emily died, it was Lavinia who searched her sancrosanct domain with instructions to burn letters, unopened packages, and manuscripts. As she was carrying this out, setting fire to pages before she had read them herself, she opened a bureau drawer that concealed an enormous hoard—the pages of Emily's lifework. Rather than the folder of light verse she might have expected, Lavinia found over 600 poems in the first box, and later many more, totally unordered and in various stages of completion. Words were scribbled in the margins; there were fragments, a confusion of differing versions, and words accompanied by an array of possible substitutes. Emily, apparently, had saved everything.

"Soon after her death her sister Lavinia came to me," wrote Mrs. Mabel Loomis Todd, a friend who would be Emily's first editor, "as usual in late evening, actually trembling with excitement. She told me she had discovered a veritable treasure—quantities of Emily's poems which she had had no instructions to destroy. She had already burned without examination hundreds of manuscripts, and letters to Emily, many of them from nationally known persons, thus, she believed, carrying out her sister's partly expressed wishes but without intelligent discrimination. Later she bitterly regretted such inordinate haste. But these poems, she told me, must be printed at once. Would I send them to some 'Printer,' and how quickly could they appear?"

After several entreaties, Mrs. Todd undertook the cumbersome, baffling, but rewarding task of sorting and editing the letters and poems. For four years she studied and edited daily, determining some chronology in the work by distinct changes in the poet's handwriting. The first volume of poems was published on November 12, 1890, and sold out six editions in five months; a second series appeared the following year, and a third in 1896.

?

How did Lauren Bacall meet Humphrey Bogart?

While moviegoers across the country flocked to the films of Humphrey Bogart to swoon over his aloof machismo or be dazzled by his daring, teenaged Betty Bacal stood in the shadows of the streets of Brooklyn, as yet unknown. Hanging out in front of Sardi's, she sold a theatrical sheet, *Actor's Cues*, and brazenly introduced herself to any passing director or producer. She'd quit the American Academy of Dramatic Arts with the intention of sailing straight into a stellar career on Broadway. But in three years she found only minor roles in three shows, and these flopped and quickly closed. One was *Johnny 2 x 4*, in which she played a bar girl in a 1926 Greenwich Village speakeasy and had no lines at all. To make ends meet, Betty modeled in the garment district and later got a job in a theater as an usherette, earning $8 a week.

Luck was with her one night at Tony's, where she was introduced to Baron Nicki de Gunzburg, an executive in the fashion department of *Harper's Bazaar*. He in turn introduced her to the magazine's top photographer, Louise Dahl-Wolfe, and fashion editor Diana Vreeland, both of whom were struck by the young actress's angular features, wonderful skin color, and long, lanky figure. The February 1943 issue featured nine shots of Betty—by now she had added another *l* to Bacall to avoid mispronunciation—and the following month she stood on the cover in fabulous vampirish garb before a Red Cross blood bank. Three thousand miles across the country, Nancy Hawks, wife of Warner Brothers director Howard Hawks and frequently cited as one of the ten best-dressed women in America, was struck by the unusual Bacall. She got her husband to call his agent, Charles Feldman, and just as Betty was about to sign with Columbia as *Bazaar*'s representative for the film *Cover Girl*, a frantic call came through from Hollywood.

Three days later Betty was out there, meeting with Hawks and preparing for her first screen test. Although this proved not entirely successful, Hawks found Betty (whom he renamed Lauren) extremely photogenic and liked, as he put it, her "cohesive physiognomy." In May 1943, she signed a seven-year contract and began studying acting and singing, but for months no work was forthcoming.

Finally, Betty was presented with the possibility of playing opposite the charismatic star Humphrey Bogart in *To Have and Have Not.* (The novel by Ernest Hemingway had originally been sold to Howard Hughes for $10,000; Hawks bought it from Hughes for $80,000, and a script, which radically departed from the book, was being prepared by Jules Furthman and William Faulkner.) But this prospect didn't thrill Bacall. "How awful to be in a picture with that mug, that illiterate," remarked Bacall, who preferred Cary Grant. "He mustn't have a brain in his head." Bogey's initial impression of her wasn't much better. The two were introduced by Hawks while Bogey was shooting *Passage to Marseilles.* ". . . She was merely a prop," said the actor, "which could add a lot or could ruin the picture."

For the next six months Bacall prepared to try out for one of the two leading female parts. She got the role of Bogey's girl and the next time the two met the brusque actor changed his tune. "Saw your test," he said. "I think we'll have fun working together, kid."

?

How did they put Al Capone in jail?

When Al Capone's armor-plated Cadillac with its bulletproof glass and tail gunner's movable back window cruised along the streets of Chicago, people spilled into the streets hoping to catch a glimpse of Scarface. Inside, the king of the underworld reclined on silk cushions and puffed on a fat cigar, while his bodyguard, armed with a Thompson submachine gun, kept watch from the front seat. Between the late '20s and the early '30s, Capone reigned supreme, his rackets extending far beyond the city limits.

Booze was the biggest stake, these being the years of Prohibition, but in addition to breweries and bars, his organizations owned or controlled warehouses, trucking companies, restaurants, nightclubs, dance halls, brothels, racetracks, and casinos. Soon, he was infiltrating unions, film production, and scores of other industries. The Internal Revenue Service estimated his fortune to be $20 million, although *he* didn't tell them that. Well over half of Chicago's police force was on Capone's payroll, and at his immediate disposal was an army of 700 storm troopers. His influence was considerable in the political arena, and Jake Lingle of the *Chicago Tribune* was undoubtedly not the only journalist amply paid for a sympathetic portrait of Capone.

The young gangster was essentially Mayor, Governor, and machine boss by his early 30s. Stories of his munificence were widespread—how he set up the first soup kitchens in Chicago in the Depression, sent $1,200 to a Philadelphia orphanage, paid hospital bills of complete strangers, gave new friends diamond belts and gold cigarette cases studded with rubies. He threw wildly extravagant parties at his estate in Florida, and a three-day bash in Chicago in honor of the Dempsey-Tunney fight had a liquor bill alone of $50,000.

In the early years of his dominance, the public seemed willing to overlook the seamier side of Al Capone, forgetting to question just whose money he was being so generous with. Capone claimed he only wanted to carry on his rackets in peace—"All I ever did was to sell beer and whiskey to our best people. All I ever did was to supply a demand that was pretty popular. Why, the very guys that make my trade good are the ones that yell loudest about me. Some of the leading judges use the stuff." Since the law forced the trade underground, it turned in on itself, breeding violence and corruption. Thus the "hero" Capone killed 20 to 60 people and was responsible by delegation for an estimated 400 deaths. Gradually, the philanthropist became the public enemy and forces gathered to put the gangster away. But such were his power, wealth, and control—all at a comfortable distance from the sordid scenes of crime—that no one could pin anything on him. And if they could, they did not dare.

The first time Capone went to jail, it served his own purposes. Having just left a convention of gang leaders in Atlantic City,

King of the underworld Al Capone enjoys his riches, taking life easy on a yacht in Biscayne Bay, Florida.

New Jersey, Capone and his sidekick "Slippery Frank" Rio stopped in Philadelphia. As they came out of a movie, two policemen picked them up for possession of concealed deadly weapons. Capone had been picked up only a few times before—in Illinois for possession of a weapon, in New York on suspicion of murder and later for disorderly conduct. Each time he was quickly discharged. Here in Philadelphia in 1929, Al expected the usual three-month sentence and pleaded guilty. But Judge John E. Walsh of the criminal division of the municipal court slapped a year's sentence on him, and a stunned Capone went off to the Holmesbury Penitentiary. Why didn't he set his elaborate legal machinery into action? The rumor was Capone wanted to escape Sicilian vengeance for three recent murders, and decided to stay out of the open for a while.

In prison, he never slept better. And the docile gangster was the model prisoner, ingenuous, responsible, and diligent. He was soon moved to more comfortable lodgings in the Eastern State Penitentiary, where he enjoyed his own cell replete with carpet, armchairs, and a $500 radio. In reward for his conduct, his sentence was commuted to ten months, at the end of which the prison went to great lengths to arrange a surreptitious departure to elude the press.

Things had not gone well for the Capone kingdom during the months he was in jail: pressure for repeal was building, business was dropping, and 26-year-old Prohibition agent Eliot Ness's special squad had smashed into six of Al's breweries with their ten-ton truck. Dubbed the Untouchables, Ness's ten-man group destroyed $25,000 in equipment and held a parade of 45 confiscated beer trucks, which threw Capone into a fury. Despite threats and attempts on Ness's life, the Untouchables continued to close liquor plants and destroy equipment. In 1931 they shut down a huge alcohol plant disguised as a paint store, which produced 20,000 gallons a day, issued directly to nearby railroad cars. Capone claimed that he wanted to retire, but he was in too deep. Anyway, his actions belied that desire, for this was when he turned his racketeering to industries, unions, and politics.

Meanwhile, the forces of law and order in Miami and Chicago were not the least bit pleased to have Capone on the streets again. Nor was President Hoover, who instructed Elmer J. Irey of the Treasury Department Enforcement Branch to open an investigation and get Capone behind bars. The recent arrest and conviction of "Bottles" Capone, Al's brother, paved the way for the attack. Eddie Waters of the Chicago Internal Revenue office was noted for putting the squeeze on gangsters to pay their taxes. He had dug up all kinds of bogus bank accounts in which Bottles was concealing funds, and eventually got him convicted on an attempt to cheat and defraud the American government. Much encouraged by this, Irey turned to Arthur P. Madden, agent in charge of the Chicago Intelligence Unit, who assigned two agents to the job: Frank Wilson (later head of the U.S. Secret Service) and an Irishman known by the pseudonym of Patrick O'Rourke. For months Wilson plowed through masses of account books confiscated from brothels, distilleries, and casinos, trying to disentangle the maze of false names and insiders' lingo. O'Rourke, meanwhile, assumed the identity of Michael Lepito, a gunman from Philadelphia and Brooklyn, and took the hair-raising step of joining the Capone mob. The Secret Six, an underground group specially appointed by the Chicago Association of Commerce and including a few prominent millionaires, supplied O'Rourke with $75,000.

In the summer of 1930 Capone and his lawyer visited Frank Wilson more than once in an effort to show the gangster's readiness to comply with the law. After lengthy discussion and tear-jerking pleas that Capone had a family to support and a mortgage to pay, the lawyer admitted that his client received one sixth of the profits from a firm that did not keep records, and that prior to 1926 he made no more than $750 a week. ". . . His taxable income for the years 1926 and 1927," continued the lawyer, "might fairly be fixed at not to exceed $26,000 and $40,000 respectively. For the years 1928 and 1929 it did not exceed $100,000." For someone who sported an 11½-carat diamond ring this seemed unlikely. Wilson's investigations, furthermore, had revealed that Capone was paying $3,000 annually to the phone company, $6,000 to hotels, and he spent over $100,000 each year on horses and clothes. At last, Wilson came across the clinching evidence in an account book of the Hawthorne betting shop, which showed to whom the winnings were paid. An ex-cashier, Leslie Shumway, estimated a profit of $587,721 over a period of 22 months. If Al was getting only one sixth, he was still getting a lot more than $750 a week. In March 1931 Shumway was brought before a secret grand jury and Capone, who was not present, was charged with tax evasion in 1924. Capone, who undoubtedly got wind of the situation, left Chicago, went to Los Angeles (where he was ordered out by the Governor), then headed for Florida. Legal processes continued and in June, no longer able to ignore them, Capone returned to Chicago. The grand jury indicted him and 68 others on 5,000 violations of the Prohibition laws. Capone was also charged with failure to pay $215,000 in income taxes between 1924 and 1929 on an income of over $1 million.

O'Rourke quickly learned that Capone had hired five gunmen to murder Wilson, Irey, and Madden. But Capone abruptly abandoned the idea and tried a gentler approach. A New York "businessman" offered Irey $1.5 million to let Capone off, but Irey turned it down.

Throughout the first weeks of June, Capone's attorneys negotiated with the federal authorities. The rumor was that a secret, unofficial agreement was made whereby Capone would plead guilty to a lesser charge and, in return, get a mere 2½ years for crimes that could put him away for some 25,000 years. The authorities preferred this short-term victory to the risk of defeat with terrified witnesses and jurors fearing for their lives should Capone be convicted.

On June 16 Capone kept his end of the deal, but when the court reconvened on July 30, Judge James H. Wilkerson took a shockingly different stance. "The defendant must understand that he cannot have an agreement in this court. . . . A plea of guilty is a full admission of guilt. . . . If the defendant expects leniency from this court he must take the witness stand and testify on what grounds he expects leniency."

Capone's attorney, Michael Aherne, countered that he and Capone "never considered that this would be an unqualified plea. . . . I am frank to say that we would not have entered those pleas unless we thought the court would accept them as made." With this the defense tried to withdraw the pleas, but it was too late. Although the court lacked sufficient proof to act on the Prohibition charges, a trial was set in October to rule on the income tax charge.

Tension and high drama surrounded the 12-day trial. Capone was continually badgered by the press. Scores of witnesses attested to the sums he spent. The defense contended that the prosecution could prove nothing except that Capone was a spendthrift. But the prosecution recalled Capone's lawyer's admission the previous summer that *some* income did exist, and evidence started piling up against him. In October 18, the jury, after a ten-hour session, found Capone guilty of tax evasion in 1925, 1926, and 1927 and of failure to file a return in 1928 and 1929. The judge sentenced Capone to 11 years in the federal penitentiary in Atlanta, fined him $50,000, and assessed another $20,000 in costs.

Eliot Ness oversaw Capone's momentous departure from Chicago. Five cars loaded with heavily armed Prohibition agents and policemen ushered him from the jail to the train station. Huge crowds gathered for a final glimpse of the notorious gangster. The press was everywhere and Capone, alternately enjoying the publicity and enraged by it, remarked, "Jeez, you'd think Mussolini was passin' through."

And then he was gone, leaving a legend, and one of the most corrupt empires the nation has ever known.

?

How did they make the first accurate map of the world?

Attempts to map the earth began thousands of years before the birth of Christ. The first known maps were made by the Babylonians and the Egyptians, who transcribed information gathered from direct measurement of the ground. Babylonian ground plans, made for the purpose of taxing property, were inscribed on tablets of baked clay and date back to around 2300 B.C. As early as 1300 B.C. the Egyptians made maps on rolls of papyrus, some of which contained accurate indications of sea, roads, and towns.

By the 4th century B.C., Greek scholars, aided by astronomical observations, arrived at the conclusion that the earth was a sphere, and at around this time the Greeks used globes to represent the world. In the 3rd century B.C. the astronomer Eratosthenes, through studies of the angle of the sun's shadow at two given points on the ground, made a remarkably accurate estimate of the circumference of the earth. The Greeks' next step was to determine the three standard lines partitioning the earth: the equator, the Tropic of Cancer, and the Tropic of Capricorn. This determination was also accomplished by observing the behavior of the sun. The equator corresponded to the equinox, when the sun rose due east and set due west and the hours of daylight and darkness were equal. The tropics corresponded to the maximum height of the sun, which in turn marked the solstices, the longest and shortest days of the year. The Greeks also conceived of the

Arctic and Antarctic circles, but these parallels were not fixed as they are today; rather, their positions were based on the relation of the observer to the horizon, and their radii increased as the observer moved away from the equator.

In the 2nd century A.D. the scholar and scientist Claudius Ptolemy incorporated the discoveries made by earlier astronomers and put together an atlas of the world on 27 sheets—10 of Europe, 12 of Asia, 4 of Africa, and 1 of the entire globe. Accompanied by text, this was titled the *Geographia*. Along with his atlas, Ptolemy introduced the concepts of longtitude and latitude. Although the atlas was far from accurate, some of it being based on mathematical errors or on travelers' accounts, Ptolemy's work set the standard for map making for the next 1,000-odd years.

Throughout the Middle Ages, Ptolemy's maps, though preserved, were for the most part forgotten and replaced by unscientific, fairy-tale-like charts of the earth's regions. But with the expansion of world trade and the first nautical explorations of the world's oceans, the need for accurate maps arose. Beginning in 1300, and for the next three centuries, Italians, Spanish, and Portuguese sailors made a number of navigational charts. Called portolan charts, after the Italian word for descriptions of the locations of ports, these maps showed the coastlines of the Mediterranean and neighboring seas, and later showed the Atlantic coast. Among the tools used in making these maps were the newly invented mariner's compass and refined versions of the astrolabe, an instrument used in Ptolemy's time. As a result of the compass's use, north began to appear regularly at the top of the map, rather than to one side or at the bottom. The astrolabe, a metal ring marked in 360° with a telescopic-sighted pointer pivoted at its center, allowed sailors to pinpoint their latitude by measuring the altitude of the sun. The portolan charts were done as individual manuscripts on parchment, the size and shape of which often determined the scale of the map.

The development of engraving in the 15th century was a significant advance in the science of cartography. Ptolemy's *Geographia*, having been copied by hand over the centuries, was first printed in 1472, and it remained, until 1570, one of the most important cartographic publications. In 1570, Abraham Ortelius, who lived and worked in Antwerp, published his *Theatrum Orbis*

Terrarum, which supplanted the *Geographia* as the authoritative map of the world.

The oldest globe still in existence today was completed in 1492 by Martin Behaim of Nuremberg. The globe was 20 inches in diameter and contained 1,100 place names. From around 1500 and on through the next two centuries, European map makers regularly produced globes in addition to two-dimensional maps.

In the 16th century Gerardus Mercator, who lived and worked in Duisburg, the Netherlands, developed a method of projecting a spherical map onto a flat sheet that showed all navigational directions as straight lines. His method of projection involved stretching the surface area of the map toward the earth's poles, and maps based on his work are still used by sailors today. One of Mercator's maps, done in 1538, names North and South America for the first time.

By the 17th century the Dutch cartographers Hendrick Hondius, Jan Jansson, and Jan Blaeu were making revisions in world maps based on many years of overseas explorations. In 1648 Blaeu completed a map of the world that was accurate in terms of coastlines, but still relied, in part, on Ptolemy's work for details of the interiors of countries. Elsewhere in Europe, in 1668, the Italian-French scientist Jean Dominique Cassini conceived of a large-scale map of the world, a circular "planisphere" 24 feet in diameter to be drawn on the floor of the Paris Observatory of the Académie Royale des Sciences. On Cassini's map meridians radiated from the center to the periphery like the spokes of a wheel, at angles of 10°. Concentric circles at intervals of 10° indicated latitude, starting with 0° at the equator and numbered in both directions.

The emergence of modern map making took place in the 18th century. The advent of nationalism brought about government-sponsored surveys, with France initiating the first topographic survey in the middle of the century. A method for determining longitude accurately was also developed in the 18th century, largely due to the efforts of John Harrison, an English clockmaker. Harrison invented a clock that would function nearly perfectly at sea; such an instrument was necessary for sailors to find their longitudinal position with regard to the earth's rotation, reflected by the time of day.

Still, by 1884 no more than 6 million square miles, less than one ninth of the land surface of the world, had been surveyed. The remaining eight ninths, containing a population of 900 million, was far from being completely charted. In October 1884 the Royal Observatory in Greenwich, England, was chosen to mark the zero meridian, and longitude was marked up to 180° to the east and west of Greenwich. The zero, or prime, meridian not only standardized navigation but also coordinated the world's clocks. But the quest for ever more minutely detailed maps, based on the most sophisticated measurements, continued.

A major breakthrough in map making occurred with the advent of photogrammetry, the charting of the earth's surface by means of aerial photography. Developed in the United States in the 1930s, photogrammetry involves the overlapping of photographs, taken from directly above, to form a stereo picture, the basis for a topographical map. Stereo coverage allows map makers to see the earth's surface as three-dimensional space. On a finished map, relief is represented by contour lines. Along with aerial photographs, survey measurements are taken on the ground to provide details of elevation and location. Exceedingly accurate topographical maps produced by the U.S. Geological Survey are usually on a scale of 1 to 24,000 and cover an area of 57 square miles, or 7.5′ by 7.5′.

Although the technology of aerial photography is known worldwide, there is at present no international agency engaged in producing a topographical map of the world. Each country does its own mapping, due in part to each country's concern for national defense.

Today the United States uses the Landsat satellite system, along with aerial photography, for the purposes of map making. Landsat is a computer-assisted system that relies on two satellites in orbit around the earth. The satellites transmit thematic pictures, the colors of which are "false"—not as they appear in nature. The colors represent various features of the earth's surface, such as roads and buildings, as well as signs of mineral resources or underground waters. Because of their distance from the earth, however, the Landsat cameras do not show detail as well as a camera carried by an airplane. An aerial camera can pick up an object as small as a manhole cover, whereas a Landsat satellite can "see" an object only as small as 100 yards wide.

So what was the first truly accurate map of the world? The answer depends on one's definition of "accurate," which in turn depends on what one is using the map for. If you are a navigator who sails on oceans or bays, you would give credit to Gerardus Mercator for making the first useful map. If you are a hiker who visits remote areas in the United States, you would think of the U.S. Geological Survey's topographical maps as the first practical charts. If you are a student in a geography class, you would probably look no further than the National Geographic Society's map of the world, with a scale of 1 to 39 million. But according to one map historian, Richard Stephenson of the Library of Congress, Ptolemy's atlas was the first accurate map of the world. "Ptolemy established place names, made regional maps as well as a map of the world," says Stephenson, "and wrote a whole treatise on the subject. The effect he had on map making lasted for hundreds of years."

?

How did they name the hot dog?

Although "wiener" and "frank" and "hot dog" are interchanged quite freely today, old-time sausage makers insist that originally Vienna's wiener and Frankfurt's frankfurter were two distinct breeds, the latter being more coarsely ground and highly seasoned. To confuse matters even more, the hot dog's antecedent is called a *wiener Wurst* in Frankfurt and a *Frankfurter* in Vienna—apparently neither city is eager to lay a claim.

Antoine Feuchtwanger of Bavaria is usually credited with introducing the frank to the shores of America, selling the sausage in St. Louis in the 1880s. At the Louisiana Exposition in 1904, he offered white cotton gloves along with the piping hot franks to prevent customers from burning their hands. The gloves were expensive, though, and most were not returned, so Feuchtwanger turned to his brother-in-law, a baker, who created long buns to fit the meat. Charles Feltman was another immigrant and early entrepreneur of the frank, opening a business on Coney Island, from which his employee Nathan Handwerker departed in 1916 to found the successful hot dog outfit Nathan's Famous, Inc.

The vendors of the shrewd concessionaire Harry M. Stevens, who sold franks at the New York Polo Grounds on cold days, lured customers by shouting, "Red hots! Get your red-hot dachshund sausages!" ("Dachshund" was a name applied to numerous German things at the turn of the century, the new German-American sausage included.) Sports humorist T. A. "Tad" Dorgan was in the press box at one particular game in 1900, and hearing the vendors inspired a new cartoon. He sketched a barking dachshund in a roll and, not knowing how to spell "dachshund," called his character a hot dog—and the name stuck.

Today the hot dog legally may contain just about any meat but dog—beef, pork, chicken, mutton, veal—up to 30 percent fat, and, if indicated on the label, up to 3.5 percent cereal. A variety of names, such as beef frank, chicken frank, or just hot dog (which usually means the presence of pork), give some clues about what's in the sausage, but no one thinks twice about it at parks and stadiums. The dog remains an all-time favorite with Americans, who now consume about 17 billion a year.

?

How did they unionize the Ford Motor Company?

No one was more vigorously opposed to unions than Henry Ford. As waves of strikes disrupted factories across the country in the late 1930s, as General Motors and Chrysler made concessions, Ford stood firm, convinced to the bitter end that unions were controlled by "predatory money interests in New York." The recalcitrant tycoon, by then in his 70s, saw himself as the boss—one who would issue orders and not negotiate, but one who took paternal pride in the well-being of his workers, assured good working conditions and decent pay, and considered that in the American spirit of individualism, each worker had to measure up or get out. These ideals had served him well for years, but in the '30s the $5 day was a thing of the past. With recessionary cutbacks, jobs were scarce, cheap labor was easy to find, wages stayed low and at times dropped, and working hours increased. Ford faced the ever more potent threat of the United Automobile Workers (UAW), headed by confident young Homer Martin, a capable leader and excellent orator, who was assisted by George Addes, Richard Frankensteen, and a former Ford employee, Walter Reuther. Trouble flowed from Washington, too, as Senator Robert Wagner of New York introduced the Wagner Act, designed to ensure all employees the right to self-organization, collective bargaining, and other activities to safeguard their interests.

For a while Ford capitalized on the public's distrust of labor and suspicions of Communist infiltration: over two thirds of those interviewed in a Gallup poll in 1937 opposed the effective and increasingly widespread sit-down strikes that had paralyzed production at General Motors. Furthermore, he determined to obstruct the union efforts by delaying actions and by inciting confusion, intimidation, and disillusion. To head the battle against the organizers Ford could not have found a more aggressively opportunistic, brazen, and fearless man than personnel chief Harry Bennett. This tough ally befriended police, FBI, and powers in the underworld; he fortified the plant with "broken bruisers, ex-baseball players, one-time football stars, and recently freed jailbirds."

There followed a reign of terror in the Ford plants, each man isolated from the next and fearful for both his job and his person, fearful even of speaking with (much less organizing) his fellow workers. Bennett's pervasive espionage both inside and outside the plant resulted in the firing of employees on the slightest evidence, and in countless arrests of union agents. Working conditions, meanwhile, worsened as employees were forced to swallow the "speedup," which drove them to exhaustion, and the "stretch-out" (more machines for each man to oversee), which wore their nerves ragged. Many complained of "Forditis" or "the shakes." Many were fired so that they might be rehired at a lower wage, and no consideration was given to seniority. The company made it a policy, in fact, not to hire men over 40, who were more apt to organize labor and push for the seniority rights outlined by union demands. Omnipresent fear of the soup lines compelled workers to submit to this pressure, although turnover was nonetheless rapid, due to sheer exhaustion.

Henry Ford's antagonism to labor unions was even fierce enough to divide and alienate his own family. From the start his son Edsel took a more moderate line, wishing to acknowledge and negotiate with the UAW. But his humiliation and impotence in the face of his father's regime left him tired, hostile, and rightfully jealous of the power allotted to Bennett. His mother, Clara, took his side, and ultimately had tremendous influence in bringing around her husband, whose powers of judgment were perhaps failing.

Bennett's terror had slowed the UAW's attempt to organize the River Rouge plant in Dearborn, Michigan, so in May 1937 Reuther, Frankensteen, Ed Hall, and others decided to take deliberate action. Having obtained permission from the Dearborn city hall to distribute their pamphlets, they arrived at the Rouge on the afternoon of the 26th. Immediately they encountered Bennett's goons, strategically posted at each entrance. Angelo Carusa, boss of a Detroit gang, two professional wrestlers, and a boxer swiftly ordered the union organizers to get out. As they turned to depart, they were met head on by another legion of guards, who attacked without a moment's hesitation and without discrimination. The men were beaten and kicked brutally, Reuther thrown down some iron stairs, Frankensteen severely wounded in the abdomen. One organizer's back was broken. But the press was on hand to record the gruesome details of the incident, which became known as the Battle of the Overpass. Millions would see the newspaper photographs, which could do Henry Ford no good in the public eye, and the photos would appear in court when the National Labor Relations Board (NLRB) filed a complaint.

Although the employees were more conservative in the Dallas plant, things were not peaceful there either. Dearborn sent Warren Worley down to Texas to organize an espionage squad of 30 to keep tabs on union activities inside the facility. Two UAW agents were beaten and another group of burly factory men armed with pistols, whips, and blackjacks was stationed outside. Scores of beatings ensued, and the police not only sanctioned the violence but often warned Worley when union agents might pay a visit.

In Kansas City, the UAW was close to a victory in 1937,

claiming 2,000 members among the 2,500 Ford workers. Ford and Bennett determined to wipe out the union stronghold, and formed a company organization, the Blue Card Union, whose officers spread rumors that the plant would be moved to St. Louis. (The following year the National Labor Relations Board charged the Blue Card Union with locking out and refusing to reinstate nearly 1,000 Ford workers, penalizing over 100 others for union activities, and in general practicing interference, restraint, and coercion.) Many fearful employees withdrew from the UAW, but the organizers pressed on, and in response Bennett shut the plant for three months. The union announced a lockout strike; the company retaliated by firing all maintenance and production workers and announcing a move to Omaha. A desperate city manager flew to Detroit to plead on behalf of Kansas City. Ford agreed to reopen the plant with the stipulation that "men who wish to work are permitted to do so without interference." Thus the police were to be on hand to break picket lines and arrest union leaders—a coup for Ford.

In the spring of 1938, Homer Martin managed to obtain some oral concessions from Ford by threatening legal action under the recently enacted Wagner Act. The company would acknowledge the UAW's right to bargain for its members, workers would be allowed to wear union buttons, and those discharged for union activities would be rehired. In return, Martin would withdraw UAW suits against Ford brought before the NLRB. Perhaps the old man, celebrating his 75th birthday with much ceremony and many tributes, was softening. But the concessions were never put into effect. Prolonged dissension within the UAW led Martin to resign in 1940 and his successors allied themselves with the Congress of Industrial Organizations (CIO). A stronger labor force resulted. The company was inundated with an unfavorable press, and several distinguished union men made it their task to inform President Roosevelt of any unfair firing of union members.

The culmination came as a harsh surprise on April Fool's Day, 1941. When the union chairman and seven others were fired at the Rouge, some 1,500 men gathered at the superintendent's office. Workers shut down three buildings immediately. When Reuther arrived on the scene, 50,000 men stood there refusing to work. An official strike was declared, and the workers marched with much excitement and drama to the UAW local headquarters on Michigan Avenue. The union men blocked all entranceways to the Rouge, preventing anyone from going to work and preventing as well the departure of the 2,500 or so loyal-to-Ford employees who remained inside. These nonstrikers tried to break the picket lines and April 2 and 3 were marked by bloody battles. Eventually, food supplies inside ran low; the men in the plant took to gambling in crap games and racing around in the new cars. The plant was absolutely chaotic.

Despite the melee, Bennett and Ford would not budge. Their pleas for government support went unanswered, however, nor did their cries about communism persuade the public, and Edsel persisted in demanding the union be recognized. Bennett, in fact, laid all blame for the outcome on his shoulders, later claiming, "I don't think the CIO would have won if it hadn't been for Edsel's attitude!"

As it was, the strike ended with Ford's concession to allow an election under the NLRB, by which the employees could choose a union to represent them in establishing a contract. The outcome of this election, held in May, showed 70 percent in support of the UAW, 27.4 percent for an AFL-affiliated union, and a meager 2.7 percent against any union at all. Negotiations for a contract took place in Detroit, Pittsburgh, and Washington, with Ford lawyer I. A. Capizzi, Harry Mack, Frank Nolan, and sometimes Bennett representing the company, and R. J. Thomas, George Addes, and Philip Murray speaking for the UAW. They prepared a contract and presented it to Ford, who had suffered a stroke and was failing rapidly. "I don't want any more of this business," he exclaimed. "Shut the plant down if necessary. Let the union take over if it wishes!" It was at this point (according to Allen Nevins and Frank Hill in their authoritative history of the Ford company) that Henry's wife Clara brought pressure to bear, even threatening to leave him if he did not sign. Nevins and Hill suggest, furthermore, that Ford had showed readiness to concede when he spoke to Martin in 1938 and that now he considered public opinion as well as the embarrassment of the sordid NLRB hearings.

He signed the contract and it was a stunningly generous one. Edsel, furthermore, said it would apply to *all* Ford employees in all the plants. Now the workers would have a union shop. All employees would belong to the UAW and the company would deduct initiation fees and monthly dues from each worker's pay. Over 4,000 workers who had been fired would receive back pay, and the company agreed to pay wages equal to the highest amount paid by any competitor, as specified by the UAW. In return, the union dropped all charges against the company. At last the battle was over. The way was open for a new method, a new spirit, and the alleviation of at least some of the grievances of the weary forces of labor.

?

How did the Indians start scalping?

By imitating the white man.

Godwin, Earl of Wessex, scalped his enemies way back in the 11th century. Six hundred years later the Dutch and English brought the custom to North America, not as an official method of warfare, but as a bounty to ease the fury of the frontiersmen. Outlaws, runaways, and freethinkers were the early settlers of the western border, an area rife with drink and disease, a hotbed of conflict both with the Indians and with the profiteering colonial government, which was staid and tyrannical in the eyes of the freethinkers. The frontiersmen turned on the Indians, eager to clear them out, and then they turned to their governments, demanding retribution for depredations caused by the Indians. The Dutch government and soon after the English devised the scalp bounty as conciliation—that is, they paid a certain amount for delivery of an Indian scalp.

Although the major task of pushing out the Indians was being accomplished by government troops, certainly the bounty encouraged resistance to their presence. In 1703 Massachusetts paid £12 for an Indian scalp. Less than 20 years later the price had soared to £100, an enticing bonus for the frontiersmen, who did not care (nor could the governments know) whether the scalp came from an Indian enemy or ally, from a man, woman, or child. The practice became widespread, and missionaries were solicitous for their Indian converts. The French used the scalp bounty to eliminate the peaceful Beothuk of Newfoundland, who presented only a slight irritation. During the French and Indian Wars, General Braddock offered his troops £200 for the scalp of the Delaware leader Shinngass—40 times the price offered for a French soldier's hair. Some governments continued to pay for Apache scalps into the 19th century, when a public outcry finally ended the practice.

Some Indian tribes had practiced scalping to a very limited extent before the Europeans arrived, but the scalp bounty induced widespread retaliation. Many tribes, the Iroquois in particular, used scalping as a defense against the encroaching whites and were later held accountable for initiating the practice since no European, it was readily presumed, would stoop so low.

?

How did Stradivari make his violins?

The violins of Antonio Stradivari are renowned throughout the world for their perfect proportions, vigorous and bright tonality, responsiveness, and remarkable flexibility in the hands of different musicians playing continually new music throughout the ages. The gifted instrument maker, who lived in the town of Cremona in northern Italy, did not, however, spring out of nowhere with his masterpieces and take the world by surprise. Rather, his work was the pinnacle of achievement of over two centuries of expert craftsmanship, experiment, and development by the Cremonese and Brescian schools.

The Hill brothers, themselves musicians, instrument makers, and authors of the most authoritative study of Stradivari, published in 1902, picture the legendary genius as a rather plodding, earnest, quiet fellow who methodically persisted in his attempts for perfection. But this continual searching throughout 70 years of professional work resulted in a stunning number and variety of violins, no one of which is exactly like another in proportion or tone. Herein lies his genius, and his departure from his teacher, Nicolò Amati, who perfected the recipe used by his family in three previous generations—albeit a magnificent one, which brought him the reputation of being the finest living Italian violin maker until his death in 1684.

Young Stradivari, born in 1644, was apprenticed to Amati at the age of 12 or 14. A violin recognized by the Hills as an early Stradivarius bears the label "*Alumnus Nicolai Amati, faciebat anno 1666,*" the customary signature of a student, and by the following year, the pupil was signing his own name. After several years of apprenticeship, he stayed on awhile in the prestigious Amati house as a paid worker, making the small, light, beautiful instruments for which his teacher was noted.

The Cremona makers were fortunate to have superb materials readily at hand: pine and spruce, whose straight fibers do not easily warp, and maple, which has a slightly higher density and transmits vibrations more slowly. Earlier makers had used cedar, poplar, lime, or pear, but maple became the favorite because it is easier to model and more beautiful, and produces a brighter tone. Although theories circulate about Stradivari's mystical understanding of the acoustical qualities of wood, the Hills maintain that the choice, while discriminating, was based on what was in stock and on what a client would pay (some offering to buy more expensive imported wood). The wood supply in Italy was ample, however, and the climate suitable for drying the wood in the sun and air for about five years, as was necessary before it could be used without fear of warping.

Working diligently in his little shop, Stradivari produced an estimated 1,116 instruments—violas, cellos, and, by far the most numerous, his celebrated violins. The method of creating these violins began with choosing a fine maple log and cutting it either "on the slab" (vertically across the grain) or "on the quarter" (in pie-shaped wedges). The former produced one flat piece, which might be used for the back. In the latter method, two pieces were often required to form the back; when joined at the thicker end, the wedges produced a mirror image symmetrical with the center line. Very slowly and carefully, Stradivari shaped the back and belly (usually composed of two pieces of pine) with saws, gouges,

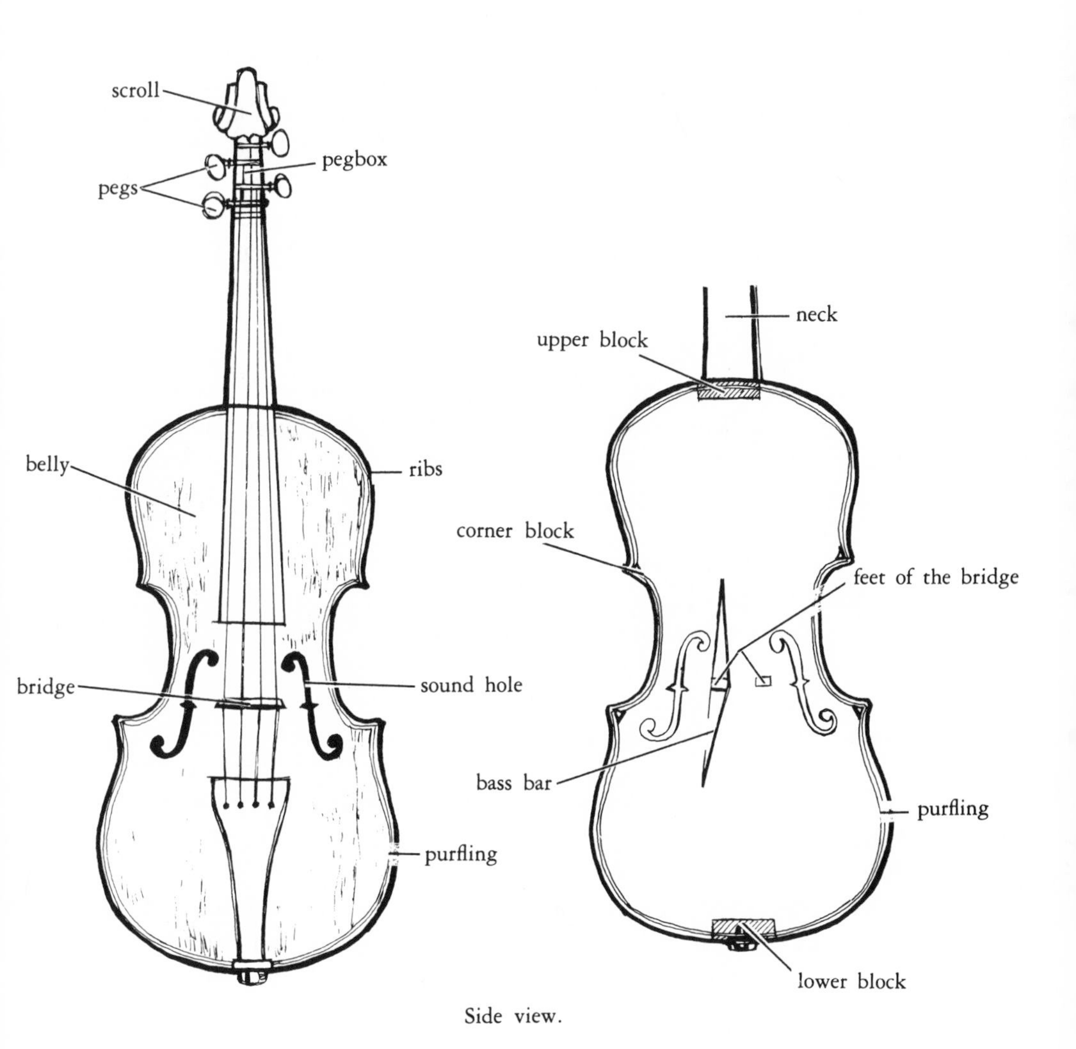

Side view.

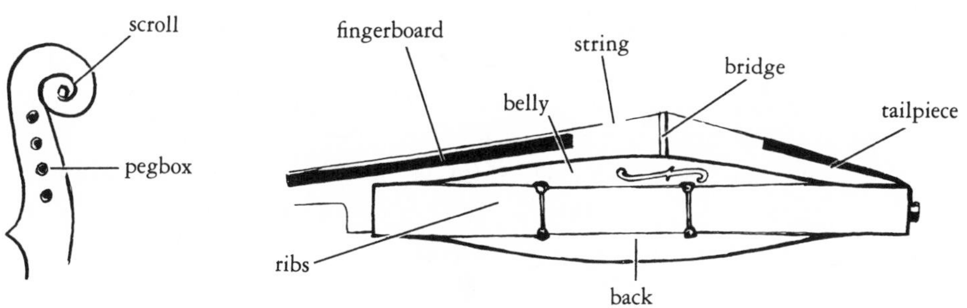

The component parts of a violin.

planes, and calipers. Thickness has a marked influence on tone and the early makers recognized the necessity of the back's being slightly thicker in the center: for example, ten sixty-fourths of an inch in the center, graduating to six sixty-fourths at the flanks. The bellies are usually about six sixty-fourths all over. Guided by intuition and a keen eye, Stradivari varied these measurements according to the quality of the wood and its curvature, which is also a significant factor, a flatter instrument producing a stronger, harder tone.

Stradivari was a superb draftsman, and he drew outlines of his instruments, from which he created molds for shaping the sides. Some of these molds are still in existence and the craftsman either had many or, more likely, was somehow able to adjust them to constantly changing dimensions.

Belly and back are held together by ribs, one twenty-fourth of an inch thick, which when wet were bent to the shape of the mold with hot bending irons. These were glued in place and further supported by wedge-shaped linings that Stradivari, unlike those before him, extended into the center blocks. Whereas Amati used willow for these linings and pine for the various blocks that strengthen corners and support the neck, Stradivari used willow, a light but strong wood, for both. Furthermore, he reduced the size of the blocks as much as possible in order to leave the sides freer to vibrate.

Once shaped, the back was glued to the maple sides and the belly followed, after the addition of bass-bar and sound holes. The former is a narrow strip of pine glued to one side of the belly in order to inhibit vibrations on that side and produce graver bass notes. In this Stradivari basically followed the example of his teacher; in the sound holes, he left the signature of his skill and ingenuity. Although meticulous in his measurements, the instrument maker was never rigid or redundant in his carvings, for each set of sound holes was perfectly suited to the arching and proportions of that instrument. These graceful holes, shaped like a cursive *f*, affect the vibration of the instrument and thus the tone,

Sound hole of the "Dolphin," 1714.

and are extremely difficult to cut because of pine's tendency to split. Using compasses, Stradivari determined the precise position of the top and bottom of the sound holes; he traced the longitudinal line with the aid of a small template, then left the form of the wings and the curve of top and bottom to his artistic eye and free hand.

Stradivari smoothed the interior of the violin with glass paper and made sure that the body, once glued, was completely airtight, as the slightest gap would create a buzzing sound. Then came the demanding task of adding the purfling, three narrow strips that follow the outline of the instrument on the front and back, decorating and helping protect the body of the violin. Stradivari cut a small groove along this line with a sharp knife, then inserted the purfling and tapped it in place with a small hammer. Modern instrument makers do this as soon as the back and belly are cut, but Stradivari waited until the body was assembled. The Hills believed this method "left the maker a free hand to correct and alter the curves . . . and Stradivari undoubtedly did so, as it is very rare to find the curves of the back and belly of any of his instruments in exact agreement: in some, indeed, a considerable difference exists." The sweep of the purfling highlights the curves of the instrument, and in varying the thickness of the purfling and its distance from the outer edge, Stradivari again left his mark of elegance and balance.

Stradivari carved the neck and head from one piece of wood. The heads are always scrolls, never figures, first traced onto the wood using templates, then carved with utmost accuracy and symmetry. The neck was glued into place and the belly temporarily removed in order to drive in several nails through the interior blocks to help hold the neck.

The instrument maker then added numerous small pieces: a smooth fingerboard of maple, sometimes veneered in ebony or inlaid with ivory purfling (a set of instruments for the Grand Duke of Tuscany bore the Medici arms in mother-of-pearl); a tailpiece to match the fingerboard; pegs and pegbox; and a bridge in any of several designs similar to those made today. No one really knows how long construction took to this point, and a vital and time-consuming step was still to come.

The composition and application of varnish on a stringed in-

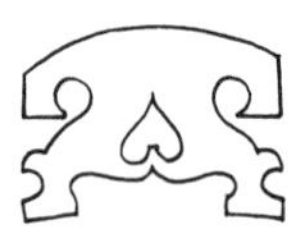

A bridge by Stradivari.

strument is extremely important. Some believe it is paramount, *the* key ingredient to the luscious tone and long life of a Stradivarius. Hard varnishes produce a metallic or glassy sound, while thick, heavy oils inhibit vibration. Somewhere in between lies the light, transparent, elastic coating of the Italian master, which gives rise to a full, woody tone. Much has been written on the subject of Stradivari's unique recipe and countless attempts have been made to duplicate it—all without success. We know only that the varnish was composed of gum, oil, and coloring ingredients, but the balance and subtle variations remain a mystery. Whatever the proportions, we know from one of two extant letters of Stradivari, in which he apologizes for a delay in delivery, that the varnish required a long time to dry. Numerous coats, possibly seven or eight, were applied and dried in the sun. Initially Stradivari produced a golden varnish in the tradition of his master, but soon moved to deeper, warmer tints of reddish orange and even brown.

A romantic story surrounds the varnish of Stradivari, which suggests that the recipe may still exist somewhere, though many experts think it unlikely. A Signor Mandelli wrote in the mid-19th century to a descendant of Stradivari, Signor Giacomo Stradivari, and beseeched him for some clue about the varnish. In reply, Giacomo confided that a "Bible, inside the cover of which was written, in the handwriting of Antonio Stradivari, the famous recipe for the varnish and the way to apply it, was destroyed. Previously, however, I made a faithful copy of the same, which I have jealously guarded. . . ." He added in a later letter, "I have never confided the secret . . . even to my wife or daughters." Giacomo wrote that he had turned down an offer by a Frenchman of 1,000 francs in order to safeguard the recipe for future Stradivaris who might once again put it to use.

Today estimates vary about the number of surviving instruments by Stradivari. Herbert K. Goodkind (*Violin Iconography of Antonio Stradivari, 1644–1737*) lists 630 violins, 15 violas, and 60 cellos. It is sad to consider how many have been lost forever, but the prizes that have been passed down through the generations, lost and recovered, brought to life in the hands of Niccolò Paganini, Nathan Milstein, Jascha Heifetz, and other virtuosos, or finally donated to museums, testify to the ceaseless innovation of their maker. Until 1690 Stradivari made violins similar in size to the Grand Amati, measuring 14 inches. These are noted for their bright sound and quick response. The beautiful "Hellier," made in 1679, one of the few inlaid violins of Stradivari, foreshadowed with its larger proportions a later development. Along the lines of the Brescian school, Stradivari began making a larger instrument, the "Long Strad," in the 1690s in order to create a darker, fuller sound. These violins have wider purfling, bolder sound holes, and varnish of a deeper color; the entire design is adjusted to meet the proportions of the body, lengthened by five sixteenths of an inch. Later, Stradivari returned to a smaller size, improving still further the sonority of his instruments, as heard in such masterpieces as the "Betts" of 1704, the "Alard" of 1715, and the "Messiah" of 1716. This last, with its velvety red finish, was made when Stradivari was 72.

Today a great many concert violinists and collectors are willing to pay considerable sums for an instrument made by Stradivari. Prices have soared to over $1 million; however, a more representative range is from $150,000 to $350,000. The condition of the violin and the period in which Stradivari made it are determining factors in the cost, with the small violins of his late period bringing a higher price than the earlier long ones. The foremost factor, of course, is the extent of the buyer's desire to own one of these treasured instruments and, one would hope, to hear it.

?

How did they determine the age of the earth?

For many years James Ussher, Archbishop of Armagh, was the authority on the age of the earth. Calculating from the genealogy of patriarchs in the Old Testament, the Irish prelate pronounced in 1648 that the world was created in 4004 B.C. So convinced was he and nearly everyone else of the divine origin of the earth that any attempt at a mechanical explanation was severely discouraged. One thus accused of impiety was the Englishman Edmund Halley, who in 1694 proposed that the earth's age could be calculated by measuring the salinity of the oceans. His theory held that this salinity increases over time because of the evaporation of water and that, if the rate of increase could be calculated, one could compute backward to the point where the seas were fresh when the earth was born. Halley did not actually come up with a figure, but he did suggest the world was a good deal older than Ussher staunchly maintained.

A century later James Hutton, a Scottish geologist, developed a theory called uniformitarianism from his observations of erosion, the gradual changes in the earth's surface caused by weathering. Believing this erosion process to be basically uniform, he rejected a cataclysmic beginning in 4004 B.C. and decided the earth had been forming over a much longer period of time.

This intuition was given a far stronger voice by Charles Darwin in the middle of the 19th century. Through the study of fossils he developed his theory of evolution, pointing out that even the slightest change in a species occurred extremely slowly over a vast period of time—longer than the memory of the human race. Millions of generations may have passed since the time of the fossils, and an enormous amount of time elapsed while prehistoric creatures themselves evolved from soft-bodied living things. Darwin suspected the earth to be thousands of millions of years old, and was very angry with Lord Kelvin, the mathematical physicist, who offered proof to the contrary. Kelvin pointed out that the earth had originally been a molten mass, that the heat at its core was gradually flowing outward at a roughly constant rate. By estimating the rate of this loss of heat and considering the present temperature, he determined the earth to be a mere 40 million years old.

The missing factor in Kelvin's theory was the effect of radioactive substances in the earth. In 1904 Lord Rutherford discovered that these substances—thorium, uranium, and potassium—emit heat. Kelvin's calculations were thus invalidated and Darwin's suspicions triumphed.

Pierre and Marie Curie had studied uranium in the earth's rocks and found that over time this substance gradually decays, emitting particles, and eventually becomes lead. The rate of this change was then determined: 50 percent of the uranium in a rock will change to lead in 4.5 billion years (this period of time is called the half-life of uranium). Knowing the rate of change, said Rutherford, one could study the ratio of radium to lead in a particular rock and compute its age. At Yale, B. B. Boltwood did just this and found high proportions of lead in some of his specimens. He judged them to be 3.3 billion years old.

Since the earth's surface is subjected to constant erosion, with huge amounts of material blown or washed out to sea each year, it seemed very likely that the rocks Boltwood and others measured might not be quite as old as the earth itself. In fact, the current estimate of the earth's age derives from extraterrestrial rocks, meteorites, which have collided with the earth and escaped the erosion at its surface. These meteorites are 4.5 billion years old and so, it is therefore presumed, are the solar system and the earth.

?

How did Mae West's secret husband like married life?

Marrying the notorious blonde with the gargantuan ego was probably the worst move Frank Wallace of Long Island ever made. Mae West was decidedly unfit for monogamy. Wallace, on the other hand, was blinded by her skirt-swishing dance numbers and her imagination in the sack. He didn't realize why she agreed to marry, and even if he had, it might not have stopped him. Mae was simply taking the advice of a friend who pointed out her chances of getting pregnant were pretty good, given her rapacious sexual appetite, which found daily outlets with actors and stagehands.

Mary Jane West, born in Brooklyn in approximately 1893 (the

date is disputed), grew up on the vaudeville stage. She sang and she outraged the neighborhood with her brazen skirt dances. Hearing that one could get pregnant from intercourse after puberty, she decided to try it out as a young girl, and this was the start of a phenomenal sex life.

Frank Wallace entered the scene when Mae was a teenager. One look at her on the stage and Frank, whose parents envisioned a sensible wife and steady job for their son, was smitten. Wallace had a flair for show biz himself and in 1909 he found himself on a bill with a *Huckleberry Finn* act featuring Mae West and Willie Hogan. Mae liked Frank's jazz dancing and contrived to do a double with him. For ten weeks she kept strictly to business. Then one night at Fox's City on 14th Street she changed her tune, and had one been in the neighborhood at 3 A.M. one might have seen the passionate duo stumble from the theater, kissing every few steps. In her autobiography, *Goodness Had Nothing to Do with It,* West claims she told Wallace their relationship was purely physical, that he didn't appeal to her finer instincts. When Wallace inquired about just what these were, Mae retorted that she didn't know yet, but she was sure she had them.

Mae consented to a secret marriage on April 11, 1911. Within a week, while the two were on tour in Minneapolis, Mae took to locking Frank in their hotel room at midnight and straggling in at 3 A.M. She wasn't doing anything wrong, insisted Mae, "just having a little fun."

In the '20s West skyrocketed to stardom, appearing both in vaudeville acts and in Broadway musicals. She decided it was time Frank went out on the road, and his luck—professional and romantic—went from bad to worse. Whenever he chanced to see his wife, generally in the company of some admirer, she snubbed him. Then one day he saw her, loaded with jewels and finery, cruising along Broadway in a fancy car. Beside her was a beefy Irishman, 15 years her senior. This Mr. Timony, a retired lawyer, had taken it upon himself to aid and abet Mae's career. Naturally this entailed freeing her of a useless husband. Secret divorce proceedings were instituted and poor Wallace, who finally saw his marriage for what it was, complied, in the process suffering an emotional collapse.

Mae just went on getting what she wanted. "My ego is

breakin' records," she'd say, and it was true enough. "My measurements are the same as Venus de Milo's, only I got arms."

?

How did they start the Chicago fire of 1871?

In the popular legend, poor old Mrs. O'Leary's cow takes the rap for the devastation of downtown Chicago, with a loss of 17,000 buildings and $200 million. The accused allegedly kicked over a lantern, which ignited the straw, the cowshed, and three and a half square miles of the city. No one else shouldered the responsibility, of course, so the quaint myth spread and persisted, providing the ruined Chicagoans with a scapegoat, an unwitting instrument of fate, at which they could only shrug and sigh.

In fact, the story was concocted for the yellow press. Years after the event, reporter Michael Ahern admitted to having fabricated it, and its success undoubtedly exceeded even his employers' expectations.

The fire apparently did start in the cowshed, but what went on there remains a mystery. One plausible account is that some friends of the O'Learys were playing cards and knocked over the lantern, or carelessly dropped the ashes of their cigars on the ground, at a time of excessive drought when fires had already been bursting out all over the city. So the record of Mrs. O'Leary's cow should be washed clean at last, while some humans are left with charges outstanding.

?

How did they first learn to make coffee?

According to Arab legend, a lone shepherd named Kaldi was roaming the hills with his herd of goats sometime around A.D. 850. One day he found his usually quiet herd behaving very strangely: old and young alike frolicked and cavorted, scampered up and down the rocky slopes bleating excitedly. The bewildered shepherd hid and watched them nibbling the berries of shrubs scattered over the hillside until, overcome with curiosity, he, too, sampled the fruit. The effects were similar. Exhilarated, he joined the dancing goats. An *imam* passing by noticed Kaldi and he too ate the berries and joined the revelry.

The *imam* returned to his monastery with the potent shrub, but neither he nor the monks could identify it. They sought advice from Mohammed. As the prayers dragged on, the *imam* dozed, and Mohammed appeared to him advising him to boil the berries of his plant in water and drink it—this, Mohammed said, would keep him awake enough to pray! The monks followed this command and called the new drink *qahwah*, which means both the invigorating and wine. Their monastery became famous for its long and lively devotions.

In Arabia, where coffee first gained popularity, this legend is the most popular explanation of its origins. The coffee plant itself, an evergreen shrub roughly 30 feet high, was brought to Yemen from Africa. Tribesmen in Kenya and Ethiopia are believed to have crushed coffee berries into a pulp, mixed them with animal fat, and carried this stimulating food with them when they traveled. Perhaps someone along the way experimented and boiled the fruit in water; although this seems highly possible, there is no evidence to confirm it.

Coffee first occurs in recorded history in a medical context. Rhazes, a doctor living in Persian Iraq in the late 9th and early 10th centuries A.D., compiled a massive encyclopedia of all known medicines, cures, and surgeries. In it he speaks of something that appears to be coffee, called *bunchum*, used for healing purposes for over a millennium. An Arabian physician at the time wrote of *bunchum:* ". . . In Summer it is by experience found to conduce to the drying of rheumes and flegmatick coughs and distillations, and the opening of obstructions, and the provocation of urin. . . . When it is dried and thoroughly boyled, it allayes the ebullition of the blood, is good against the small poxe and measles, the bloudy pimples; yet causeth vertiginous headheach, and maketh lean much, occasioneth waking, . . . asswageth lust, and sometimes breeds melancholly. He that would drink it . . . let him use much sweat-meates with it and oyle of pistaccioes and butter. Some drink it with milk, but it is an error, and such as may bring in danger of Distempers."

In Arabia the first coffee was derived from boiling fresh leaves and berries, later dried husks and seeds, of the coffee plant. The drink rapidly gained popularity for its stimulating effect, and the efforts of holy men to restrict its use to religious ritual proved futile. The *imams* of Yemen themselves took coffee on their pilgrimages to Mecca, from which it dispersed to Cairo, Damascus, Baghdad, and the rest of the Arab world. The first coffeehouses sprang up around 1475, attracting students, artisans, and travelers—a popular distraction that further annoyed the leaders of religion. During the 16th-century Muslim expansion, coffee went along, west toward Europe, east toward India.

By now a more refined brewing method was employed: the roasted beans, pounded to a fine dust, were boiled and the liquid strained through silk. Stored in earthen pots, the coffee was later reheated and enhanced with cinnamon sticks and cloves.

The Turks who invaded Arabia and North Africa under the Ottoman emperors loved coffee and it became an essential part of their social lives. Even the marriage ceremony dictated that the husband must never leave his wife without coffee beans.

Vendors were selling coffee from door to door in early-17th-century Venice. The first coffeehouse opened in London in 1652, and within 70 years, 2,000 more followed suit in that city alone. However, English women soon grew annoyed at their husbands whiling away the hours in idle talk. They voiced their complaints in a tract, *The Women's Petition Against Coffee, representing to public consideration the grand inconveniences accruing to their Sex from the excessive use of the drying and enfeebling Liquor,* asserting that coffee made them as "unfruitful as the deserts where that unhappy berry is said to be bought."

Until the end of the 17th century almost all coffee was obtained from Yemen. Java and other Indonesian islands began cultivation in 1696. Coffee growing soon spread to the West: to Haiti, Santo Domingo, Martinique, Jamaica, and Brazil in the first half of the 18th century; to Puerto Rico, Venezuela, Mexico, and Hawaii a bit later. Today the Western Hemisphere surpasses the Eastern in coffee production, and Americans consume about 138 billion cups a year.

?

How did Napoleon escape from Elba?

Napoleon abdicated on April 6, 1814, and the Allies attempted to determine his fate, according to the Treaty of Fontainebleau, signed six days later. The Emperor (who might still call himself that) was to assume sovereignty not over an empire but over the little island of Elba, lying between Corsica and the western coast of Italy. There he could sit idle with a yearly pension of 2 million francs and a bodyguard of 400, ample property, and internal affairs to oversee.

Idleness, though, was never a consideration or a temptation. The 46-year-old general launched a large-scale new government, perhaps to impress the islanders, more likely to lead the vigilant Allies to believe he intended to stay put. He rode tirelessly over the island, often starting at 5 A.M., to oversee the building of new bridges and roads, improvements in hygiene, communications, and agriculture. His nine months of exile were divided among the lovely Mulini palace overlooking the sea at Portoferraio (in which the flour warehouse was converted into a ballroom), a sumptuous country estate among the vineyards in the Valley of San Martino, and a mountain hermitage for the summer months. Nevertheless, Napoleon missed the activity of Paris, the beautiful and admiring women, and the glory of conquest. Retirement was premature; he relinquished neither his ambition nor his patriotism. He was probably conspiring to return to the mainland from the moment he left it. While on Elba, Napoleon heard that the Allies favored his deportation to some more distant island; the promised 2 million francs did not appear; the King of Naples reported that the Italian people supported him—all this propelled the Emperor to act.

Napoleon had built up a small "army" around his nucleus of Old Guard members, adding recruits from Elba and Corsica and over 100 Polish cavalry and lancers. He oversaw a decrepit fleet of ships—the *Inconstant,* on which he had arrived, a postal boat, a chebec for carrying grain, a felucca to chase pirates, and a few more. The French were suspicious, of course, and the nearby town of Leghorn on the Italian mainland was a hotbed of espionage and counterespionage.

A number of chance, unrelated events during the winter of 1815 seemed to point the way for Napoleon's escape. On January 12 a gale ripped the *Inconstant* from its moorings and drove it ashore, conveniently near Portoferraio, where repairs were begun. February 15 brought a mysterious sailor from the Gulf of Spezia who turned out to be a spy from Napoleon's former minister M. Maret, Duc de Bassano, with news that the public desired his return. On February 17, Colonel Sir Neil Campbell, the Allied Commissioner for England who kept a wary eye on Napoleon, sailed on the *Partridge* to the mainland to see a doctor—or, say some sources, to rendezvous with his mistress, the urbane Contessa Miniaci. Immediately, Napoleon issued orders to ready his battered *Inconstant:* ". . . Have her copper bottom inspected, her leaks stopped, her hull careened, and everything done to make her seaworthy," he wrote to General Drouot. "Have her painted (i.e. with black port-lids) like an English brig. . . . The brig must be re-armed, and stored with biscuit, rice, vegetables, cheese, drink—rations, half of brandy and half of wine, and three months' water supply for 120 men." When the *Partridge* loomed up on the horizon on February 24, the Emperor's plans appeared thwarted, but only English tourists disembarked. Once it was ascertained that Sir Hugh was not on board, Napoleon made little effort to conceal his actions.

On February 26 crowds rushed through the streets of Portoferraio to the harborfront to watch the Emperor depart, and at 8 P.M. a cannon fired from the *Inconstant* announced that Napoleon was on board. Then a flotilla of six vessels carrying 1,050 soldiers and gendarmes and 100 civilians sailed quietly away from Elba with a soft wind behind them. Napoleon scanned the horizon beneath the bright moon and a sky of stars. Only 50 miles away the *Partridge* sat becalmed, waiting for Campbell. And a mere 30 miles northwest of Elba lurked three French ships, one of which, the *Zéphyr*, intercepted the *Inconstant*. Fortunately for Napoleon, his Captain Taillade was friendly with *Zéphyr*'s captain and blithely told him they were making a routine run to Genoa. According to J. M. Thompson, this captain had not been given specific orders, and later claimed to have known the Emperor was on board.

Campbell eventually returned to Elba on the 28th, learned of the escape, and attempted a hot pursuit—only to be becalmed. Meanwhile, Napoleon stopped at Antibes, camped out under the stars, and enjoyed a welcoming tune by the band of the district's Governor, who was out for a picnic. At Cannes an angry butcher decided to do Napoleon in with a musket but the townsmen dissuaded him. On March 2, the Emperor headed over the mountain road to Grenoble. There, legend has it, Napoleon thrust his naked breast toward some threatening troops, exclaiming, "Soldiers, if there is one among you who wishes to kill his Emperor, he can do so: here I am!"

All along the way, for 18 days, Napoleon appealed to the French people and to the Army with his old bravado and fire. ". . . We have not been beaten," he declared, "but betrayed. . . . Shoulder once more the standards that you carried at Ulm, at Austerlitz . . . at Montmirail. . . ." By the time Napoleon reached Fontainebleau, Louis XVIII had fled his capital.

?

How did Louisa May Alcott write *Little Women*?

Much against her wishes. She hated little girls and wrote her bestseller for money.

?

How did men decide to wear neckties?

In this century convention appears to have the upper hand with men's clothes. The standard, acceptable uniform is dutifully worn by legions of businessmen pouring into offices five days a week—despite the buttoned collar that grows smaller with passing hours, despite the choking necktie cramping one's style. For some, fumbling with a necktie at 7 A.M. is the bane of existence, far removed from the practical and, then, *decorative* origins of style.

Roman legionnaires were the first to wear neckbands and for them the cloth served a purpose: to keep them warm and absorb the sweat of the neck. Louis XIV's soldiers also wore neckbands

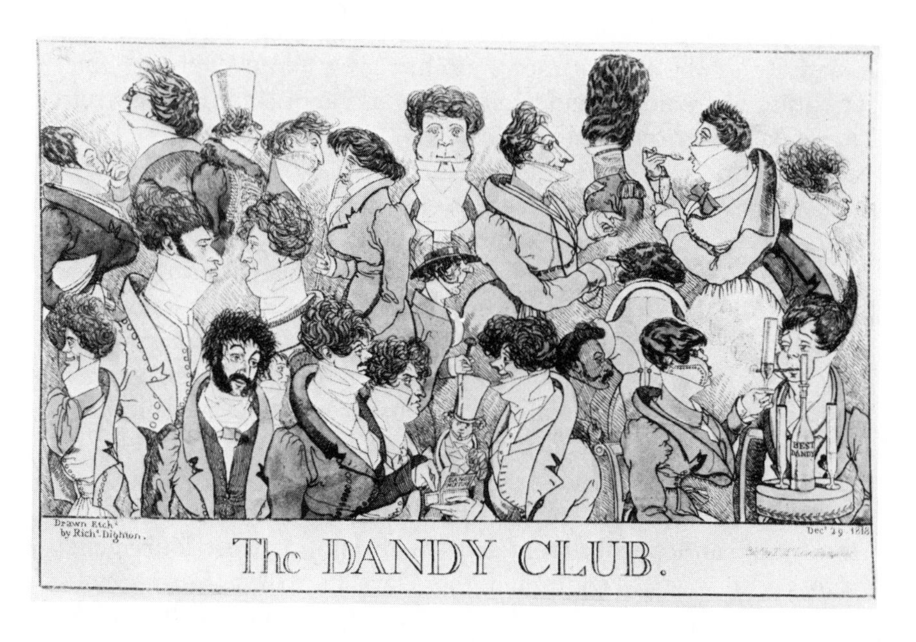

The height of fashion and discomfort. Gentlemen sport throttling cravats in The Dandy Club, 1818, *by Richard Dighton.*

and from his Croatian regiment—called Cravates by mistake in the 17th century—a name was taken for the garment. In 1692 women lined the streets of Paris to hail the return of the victorious Duc de Luxembourg, the Prince de Condé, and others from the Battle of Steenkerke in Belgium and took note of their gallant cravats. "At this time," explains Voltaire, "men wore lace cravats which required some time and patience to put on. In their haste to prepare for battle, the princes had carefully twisted these cravats round their necks; women now began to wear an article on the same model called *Steinkerque* or *Steinkerk*."

The 17th century saw a variety of cravats, in muslin and lace with knotted ends. Men were also at liberty to select a stock, a piece of cloth placed around the neck and buckled in the back, similar in appearance to the neckband worn with a riding habit today. Gentlemen in the 18th century suffered with stocks of increasing heights and stiffness.

At the time of the French Revolution, the neckband could signify one's political stance: white for the conventional, black for the revolutionaries. Increasingly in the 19th century, the cravat came to embody the essence of the man, a last vestige of creativity as the rest of men's clothes grew more uniform. Such attention was paid the cravat, its form and knot, that Balzac wrote a manual describing the intricate methods of fastening it. From Germany in 1826 came a little book titled *Cravatiana oder neueste Halstuch-Toilette für Herren,* satirizing the more serious intentions of London and Paris writers. Of a certain style labeled the *cravate à la Byron,* the book's author wrote, "Everything that issued from Lord Byron's eccentric genius, bears the stamp of a certain originality. Hence we cannot expect, in the manner of cravats assumed by the famous poet, either the studied nicety nor the petty fussiness generally to be observed in the neckcloths of the fashionable. . . . The bard of the corsaire has thrown off all restraints." Byron's tie was wound around the neck only once, starting from the back (rather than the front, as was customary at the time), and tied in front in a huge knot at least four inches across. "Its color," continued the German writer, "in harmony with the sombre flight of fancy of its originator, is that of the scabious."

The gentleman who stands in the doorway of Édouard Manet's

The Balcony wears a tie with long, loose ends, an early example of what would become the conventional style in the late 19th and 20th centuries. Today manifestations of personality have been reduced from the method of tying to the pattern on the tie. Some sport such emblems of clubbiness as horsehoes and fishing flies, tennis rackets and sailboats. There are stripes for the preppie, reserved circles and diamonds for the "old boys"—even, as advertised in *The New Yorker*, a tie with wasps.

?

How did Beethoven compose when he was deaf?

Beethoven noted the first signs of deafness as early as 1801, when he was only 28, with numerous piano and chamber works behind him but only one symphony yet created. "Let me tell you," wrote the young composer to Karl Friedrich Amenda, "that my most prized possession, *my hearing*, has greatly deteriorated." He commented parenthetically, "When I am playing and composing my affliction still hampers me least; it affects me most when I am in company," then pronounced, "Why, at the moment I feel equal to anything. . . . I have been composing all types of music, except operas and sacred works."

Beethoven closed this letter with a plea for secrecy about his condition. Some remained unaware of the problem, believing that when the master failed to socialize, for instance, or to reply when addressed, he was merely distracted—and that, too, may have been the case, particularly in company he did not relish.

If Beethoven's composing was not severely affected by his hearing failure, his public life as a pianist and conductor certainly was, and this was devastating. "Oh, if only I could be rid of it [his affliction] I would embrace the whole world," Beethoven wrote to his close friend Franz Gerhard Wegeler in 1801. "I will seize Fate by the throat; it shall certainly not bend and crush me completely. . . ."

Beethoven continued his public concerts for 13 years, for he was not, as many believe, stone-deaf right away. He even courageously undertook the direction of his opera *Fidelio* at the Josephstadt Theater in 1822 when he could hear nothing, but this attempt proved a disaster when musicians and director got hopelessly out of synch, and the humiliated composer, once informed of the situation, fled in the midst of the performance. "In the long years of my association with the mighty composer," wrote Beethoven's intimate friend and biographer Anton Felix Schindler, "there was never any experience to equal that day. . . . He never wholly recovered from the effect of this blow."

Nevertheless, the years in which Beethoven's hearing gradually worsened were ones of enormous creativity. It has even been proposed that his illness enhanced his work, forcing him to retreat from the public eye, allowing more time for composing, and enriching his soul and in turn his work through suffering—a popular Romantic notion. This, of course, is highly speculative. With a genius of the stature of Beethoven, it does not seem likely that outward circumstances would have hindered his artistic development. "The nine symphonies would have been composed had

The brooding composer at work in his study in a portrait by Paul Sonntag Berlin.

Beethoven been surrounded by a wife and six rosy children, had he been as worldly as Goethe or as healthy as Verdi," believes another biographer, George Marek.

It is obviously impossible to say how Beethoven composed his masterpieces, whether in prime health or deaf. But Schindler provides glimpses, at least, of his routine and methods. "Beethoven rose every morning the year round at dawn and went directly to his desk. There he would work until two or three o'clock, his habitual dinner hour. In the course of the morning he would usually go out of doors once or twice, but would continue to work as he walked. These walks would seldom last more than an hour, and may be compared to a bee's excursions to gather honey." In other words Beethoven, like many other composers, wrote at his desk the music he heard in his own head. "While working on a score . . . he played no instrument, and because it annoyed him to see anyone go through a work that was still incomplete, even those living in the same house heard nothing of a new symphonic work until it was rehearsed." An exception to this rule was the composition of music for the piano. Then "the master would often go to the instrument and try certain passages, especially those that might present difficulties in performance. At such times he was totally oblivious of anyone present." It is perhaps for this reason that Beethoven asked the piano manufacturer Streicher in 1817 to adjust a piano "as loud as possible" for his "weakened hearing." It is possible that Beethoven was able, in a sense, to "hear" sound through vibrations. He may have touched the strings of the piano with a wooden stick (now in Beethovenhaus in Bonn, West Germany) clenched between his teeth. Schindler reports that another piano manufacturer, Conrad Graf, made a sounding board that when placed on the piano helped conduct the sound.

Beethoven had a habit of improvising at twilight, either at the piano or on the violin. "We need not say what this playing sounded like," writes Schindler, "for his external senses were incapable of guiding him; for the other people in the house his playing, especially of the string instruments which he was unable to tune, was agony to the ears. His extemporizing on the piano was seldom intelligible, for it was usually extremely agitated . . . the left hand being spread wide and laid upon the keyboard so heavily that the noise would drown the much softer playing of the right hand." This is a sad and rather eerie spectacle to contemplate: the extreme isolation of the composer, and the uncanny discrepancy between the jarring sounds he produced and the magnificent sounds he alone must have heard.

?

How did they build the Great Wall of China?

The largest man-made construction on earth—and, it is often said, the only one visible from the moon—the Great Wall of China began as many different walls linking already existing fortifications and watchtowers. During the period of the Warring Kingdoms (c. 500–221 B.C.), myriad principalities were subsumed under huge kingdoms, which inevitably fought and erected protective walls between each other. Walls surrounded the towns and may have been extended along with a king's conquests. To the north they increasingly became a demarcation between a settled agricultural society and a more mobile, nomadic one.

In 221 B.C. the immensely powerful ruler of the Qin kingdom conquered the last opposing state and united China for the first time. Qin Shi huangdi, "the First Emperor," instituted a centralized, authoritarian government, the likes of which China had never before seen. He standardized China's written language, currency, and weights and measures. He destroyed the fortifications of the other kingdoms, melted down their weapons, and built roads 50 paces wide along which he and his magnificent entourage could parade. China was divided into 36 administrative zones, and all travel was monitored by police. The new strength and unity did not come without sacrifice, however. Any criticism of the state and all philosophical discussion were silenced, and the Emperor burned all but the most utilitiarian writings or locked them away in the imperial library. He sent 460 scholars to their deaths.

Having established internal order, the Emperor turned his attention to the external borders where barbarians and nomads were a constant threat. He dispatched the already renowned general Meng Tian in command of 300,000 men to ward off the barbarians in the north. After repelling the Xiongnu (Huns), Meng Tian stayed on the frontier and oversaw the construction of the Great Wall along the Yellow River. His troops were joined by the Emperor's ample supply of convicts, as well as 500,000 peasants. When a conspiracy later arose against Meng Tian and he was condemned to die, he allegedly confessed, "Yes, I did commit a crime, and I must die for it. From Lintao all the way to the east of the Liao River, ramparts and moats stretch for more than 10,000 *li*. Surely in that expanse of land I must have severed a vein of earth. That is my crime." Although the modern observer might look instead to the human lives that were severed, this is a noteworthy instance of belief in the quasi-divinity of the earth, a belief that would gain significance in China in later years.

The principal materials used in this early construction were earth, stone, timber, and tile. Local sources were tapped, and so the composition of the wall shifted with the topography. On mountains stones were quarried, while on the desert the wall grew out of sand and pebbles, tamarisk twigs and reeds. Earthen walls were made by first affixing posts and boards along both sides and then ramming in layer upon layer of yellow earth with wooden hammers. Workers carried heavy loads on their backs or with carrying poles. Sometimes they passed materials from one person to the next along a long line from the source to the building site.

The succeeding Han dynasty (206 B.C.–A.D. 220) led a period of expansion north, south, east, and particularly west, where they opened a new trade route. The wall was extended along this area of commercial conquest, but later dynasties were unable to maintain it.

Records indicate that, over the centuries that followed, the Great Wall remained a major concern of the emperors, who intermittently drew up huge labor forces to add to and repair it. For example, in A.D. 446, more than 300,000 laborers were press-ganged to construct the stretch of the wall south of present-day Datong. The Emperor Tian Bao recruited 1.8 million peasants in 555, while Da Ye organized over 1 million workers in 607 and another 200,000 in 608. Only with the advent of the Tang dynasty in 618 and a period of great stability did work cease. The Tangs, in fact, scorned the tradition of wall building, endorsing instead more aggressive, offensive attacks to protect China's borders.

In the wake of Mongolian occupation, however, the Ming dynasty (1368–1644) wholeheartedly resumed the age-old tradition. This was a period of massive building. Hundreds of watchtowers were erected and stronger, larger walls rose to replace or add to the old ones. By this time, roughly 1,500 years after Meng Tian's time, the Chinese had bricks, which became a principal building material, and lime. Kilns fueled by timber cut locally or carried from afar were often made on the spot to produce bricks and tiles. If brought from nearby counties, the bricks were toted in baskets on the backs of donkeys or tied to the horns of nimble goats. Materials were also transported in wheelbarrows, a Chinese invention dating back to the 1st century.

Legend has it that a dragon marked out the tortuous path of the wall and the workers traced his footsteps over mountains and valleys. Another story tells that the builders initially worked out the design and construction process so precisely that when work was completed, they had only one brick left over. It remains to this day on one of the watchtowers. Massive organization and close supervision were indeed integral to the building of so huge and stolid a wall. In the Ming dynasty, commanding generals in nine towns oversaw the construction. Planners tried always to use the land to its best strategic advantage, placing fortresses in mountain passes, for instance, or in the sharp twist of a river. The wall accentuated natural barriers, such as mountain ridges; beacon towers were placed prominently and within a certain designated distance from one another. When building began, workers first had to level the land and then lay a bedrock of parallel stone slabs. They erected inner and outer walls of stone slabs, then crammed the interior with stones, cobbles, lime, and earth, and laid bricks along the top.

The result of this Herculean effort was a wall so massive that the materials that comprise it could circle the equator, even if stacked six feet high. Standing 8.5 meters high and 5.7 meters across, the wall can accommodate five horses or ten people walking abreast. It incorporates thousands of watchtowers, military stations, garrisons, and gateways. The two-story towers harbored grain and arms below, while upstairs soldiers watched for enemies. From beacon towers on the wall and nearby mountains, soldiers at nearly any distant spot could send messages rapidly to the capital, using smoke signals during the day, fire by night. A million-troop force reportedly manned the wall during the Ming dynasty.

However awesome the Great Wall appears to the western eye today, it is for many Chinese a graphic symbol of hardship and suppression. Many legends tell of the sorrows that the people bore. Meng Jiangnü was among the thousands of wives left alone when the men were called to labor. After years of separation from her beloved husband, she set out to find him, trudging mile after mile beside the Great Wall. According to the story, she at last arrived at Shanhaiguan Pass, only to learn that her husband was dead. She cried so many tears that a section of the wall crumbled, revealing the remains of her husband buried beneath. In despair, she drowned herself in the sea. A commemorative tablet at once arose from the waves, and today a temple in her memory stands at Shanhaiguan.

In the west, the monumental wall begins at Jiayuguan Pass (the Jade Gate) beneath the snow-peaked Qilian Mountains in Gansu province. For traders venturing westward, the pass marked the last outpost of civilization and security. From there the wall runs eastward, skirting the Gobi Desert, snaking along the Yellow River for over 1,000 miles, spanning grasslands and plateaus, and cresting the ridges of the Yanshan Mountains, a total of 3,720 miles to Shanhaiguan, the First Pass on Earth. The last stretch, from Bejing to the Gulf of Chihli on the Yellow Sea, is the best preserved and most frequently visited.

The idea that the Great Wall is visible from the moon has been circulating for some 60 years, cropping up in both guidebooks and scholarly works. But Arthur N. Waldron of Princeton University's East Asian Studies Department disagrees. In a letter to the editor of *The New York Times* on March 20, 1983, Waldron debunked the sensational claim, asserting that it simply is not true: "No communication with manned spacecraft makes mention of the Wall, and in any case, a few calculations demonstrate the idea's absurdity. According to my friend Dr. Alta Walker of the

U.S. Geographical Survey, seeing the Wall from the Moon is like seeing a popsickle stick from 384 kilometers."

?

How did Gutenberg print the Bible?

Fifteenth-century Europe, at last stirring from the sleep of the Middle Ages, was a place of exploration and invention. Universities opened their doors; drama and poetry flourished; a wave of humanism peaked in Italy and spread northward; Prince Henry the Navigator and Christopher Columbus set sail for new shores. In Germany, the city of Mainz became a vital commercial center where industries prospered and ships along the Rhine brought foreign diplomats and trade. This was the home of Johann Gensfleisch zum Gutenberg, who, born c. 1397 to a patrician family, was expected to enjoy a respectable life of leisure. But he was a man with an eye toward the future. He became among other things a skilled craftsman, with a talent for technology. Gutenberg saw the need for a new way to duplicate manuscripts, a replacement for the painstakingly slow method of copying by hand, carried out by monastic scribes and more recently by students as well. In modern terms, he wanted to cut costs, speed up production, and reduce errors.

Gutenberg was not, as many history books suggest, alone in this desire. Others attempted to produce mechanical or "artificial" script, and a number of towns later laid claims as the prestigious birthplace of the printed word. Notable among Gutenberg's predecessors was Laurens Janszoon Coster of Haarlem, a verger in a church. When Coster was instructing his grandchildren in reading and writing, it occurred to him that tangible aids would be useful in this endeavor, and so he cut single letters from wood. The idea of making complete pages of type followed, and Coster tried to cast his letters in tin and lead. But legend has it that on Christmas, 1440, a pupil from Mainz stole his entire stock of type. Whatever the achievements of Coster, certainly a method of wood-block printing was employed in the early 15th century for the manufacture of cards, devotional images, and printed cloth.

The mastery of craftsmanship evidenced in the Gutenberg Bible would indicate the presence of other early experimenters, but nothing has been found that comes close to that Bible in precision or magnitude. Gutenberg himself spent 20 years experimenting, synthesizing methods, and resolving problems. He carried on secret metallurgical studies in Strassburg in 1436, returned to Mainz between 1444 and 1448, and plunged into debt to establish his workshop and train workers, laboring tirelessly to produce not only his masterpiece but a printing method that would survive unrivaled for 350 years.

The underlying genius of his method was movable type: each letter of the alphabet appears on an individual piece of metal, which can be used innumerable times in different combinations on different pages of a work, and again on later works. Gutenberg's type was modeled on the most beautiful German Gothic manuscripts of his time. His astonishingly elaborate font comprised 290 characters: letters, punctuation, and the many ligatures and abbreviations required by late Medieval Latin, in which his Bible appeared.

Gutenberg first cut a mirror image of each letter onto a block of hard metal, the head of a steel punch. As a former goldsmith, he was skilled at metalwork and familiar with the tools. The letter stamp was then struck into soft metal, a small copper block, thereby creating a depressed form of the letter, which was set in a mold. Into the mold Gutenberg ladled an alloy of lead, antimony, and tin heated to 572°F, a material that melts easily and could be further processed if necessary. Gutenberg cast hundreds of letters from each stamp, then arranged them in wooden cases, capitals in one, small letters in another.

Next came the task of setting the type by hand in perfect lines, each letter straight and placed with an appropriate, aesthetically pleasing space before and after. Each page, with its 2 justified columns of 42 lines each, consisted of a staggering 2,500 individual pieces. These were bound together with string and the whole further secured by a frame on the press. Gutenberg's pressmen probably set 20 to 40 pages of text, printed them, then dismantled the characters and returned them to their boxes, only to begin all over again.

Once a page was ready for the press, Gutenberg applied to the type an ink made from pine shavings and soot. He moistened the paper and spread it over the type, then pushed the two together under the press. Unfortunately, there are no detailed descriptions of this early handpress, built by a cabinet maker named Konrad Saspoch. We know simply that a heavy iron plate was lowered, under controlled pressure, against the type form, and the ink thus transmitted to the paper.

It took three years of constant printing to complete the famous Bible, which appeared in 1455 in two volumes, with 1,284 pages overall. Gutenberg reportedly printed 200 Bibles, of which 47 survive. As well as evidence of a historic feat of technology, each book is a work of art, with its finely sculpted type, hand-painted initial letters of each chapter, and decorative flower tendrils imitative of the finest illuminated manuscripts. Gutenberg could not enjoy the glory of his accomplishment, however, nor the fruits of his success. In 1455 financier Johann Fust, from whom Gutenberg had borrowed heavily, took over the inventor's lifework—his press, his tools, even his finest workers. Little is known of Gutenberg's fate thereafter. Some say he went blind. But Fust, along with Gutenberg's former associate Peter Schöffer, became a famous printer, reaping the benefits of Gutenberg's labor and brilliance.

?

How did they develop the parachute?

Although the principle and construction of the parachute might seem pretty elementary in these days of advanced aerodynamic technology, early experiments with the devices were feats of considerable daring, usually performed before breathless crowds at country fairs. Louis Sébastien Lenormand of France was the first to demonstrate his rustic canopy parachute in action when, in 1783, he leaped from a high tower to a safe landing. Fourteen years later an excited crowd of Parisians gathered in the Parc Monceau to watch André Jacques Garnerin rise to an altitude of 3,000 feet in the basket of a balloon, abruptly cut the suspension cord, and descend gracefully to earth with a parachute resembling an oversized umbrella. Garnerin went on to exhibit his jumps at ever increasing heights in various countries, leaping 8,000 feet in England in 1802.

These daredevils may have dazzled audiences with their novel contraptions, but the idea of the parachute was far from new—Leonardo da Vinci was contemplating the principle three centuries before. He sketched a cloth pyramid that would serve to break one's fall, and he is said to have discussed these aerodynamic factors as early as 1514. For a while the idea remained in the imagination of writers: in 1595 a book by Fausto Veranzio appeared containing an illustration of a man falling from a tower beneath a rectangular "parachute" made of cloth over a wooden frame. The hero of Desmarets de Saint-Sorlin's novel *Ariane* (1632) makes a valiant escape from a prison tower with the aid of an opened-up bedsheet.

The first parachutes were of canvas, later replaced by silk, and then nylon, which is used today. An important development, undoubtedly not considered in the early parachutes, is the use of numerous different pieces of fabric, cut, sewn, and arranged to confine a possible tear to the section in which it starts. The apparatus (parachute and suspension lines) is assembled in a pack; the ripcord that releases the parachute may be operated manually, or automatically via a timing device, or by a line attached to an aircraft. The first man to jump from a plane was Captain Albert Berry of the U.S. Army in 1912. Enormous parachutes helped the Apollo 14 space capsule land in 1971. Perhaps the most extraordinary parachute is that used for getting out of an aircraft moving at supersonic speeds. In this case, the entire seat goes along—pilot, seat, and parachute are ejected by a rocket charge. Eventually, the seat is left behind and the parachute opens automatically, cutting the pilot's speed as he descends.

?

How did Harpo learn to play the harp?

When the five Marx brothers and their parents, grandparents, and miscellaneous relatives lived crammed together in a tenement on East 93rd Street in New York, it was Minnie, the boys' mother, who was always plotting and scheming to get them onto the stage. A legend in show business, she had the energy and inspiration to launch the kids in a vaudeville act that would tour the country and finally hit Broadway. But in the early days everyone was hungry. Their father, Frenchie, was never guaranteed pay for the men's suits he made at home on the kitchen table.

Groucho had already had a taste of show business as a singer, and then decided he wanted to write. Chico had selected the vocation of professional gambler when Minnie hired a large Viennese lady with a moustache to teach him piano. Chico was supposed to teach Harpo, but this cut into his time in the pool halls, so Harpo, then considered the least talented of the brothers, was left to his own resources. In fact, he never made it through his second try at second grade and learned to read from signs in the street.

Harpo was plunking out "Waltz Me Around Again, Willie" on the piano with one finger at the age of about eight, but in the corner of his grandfather's room stood an old harp, warped, half-size, and without a string. Harpo's "Grossmutter Fanny," now dead, used to play that harp in the spas and music halls of Germany, as people danced. "All that remained of its old luster were a few flakes of golden dandruff," Harpo recalled in his autobiography. "But to me it was a thing of beauty. I tried to imagine what it must have sounded like when Grandma played it, but I couldn't. I had never heard anybody play a harp."

Fifteen years passed before Harpo plucked a string, during which time he worked as a pie sorter, cigarette boy, ragpicker, delivery boy, and piano player in a rowdy bar with a brothel upstairs and in a stuffy nickelodeon parlor on 34th Street. It was one day when the brothers—Harpo included—were out on the road,

The inimitable Marx Brothers in top form. From left: *Harpo, Chico, and Groucho.*

struggling to survive with their singing act, that the indefatigable Minnie decided the act needed more class and shipped Harpo a $45 harp. The novel instrument joined the act immediately, and "after a year of hunt and pick, ponder and pluck, and trial and error," said Harpo, "I played my first solo on the harp—'Annie Laurie.'" This was in Gadsden, Alabama. "I got a big hand and a demand for an encore. The only encore I could think of was doing 'Annie Laurie' over again, with fancy long swoops on the strings (I didn't know yet that these were called glissandos) between phrases of the melody."

Thus the harpist who enchants us, who pours forth the Mozart C Major Sonata before the mirrors of an 18th-century salon in *The Big Store,* taught himself everything, and never read a note of music. At first Harpo didn't even know how to tune his instrument, and only when he spotted a picture of an angel with a harp in a Woolworth's window in Missouri did he discover he was resting the harp on the wrong shoulder. But "the presence of the harp (the harp alone, and not the harpist)," explained the musician modestly, ". . . raised our average monthly income by five dollars," so he continued to learn.

After eight years of professional playing Harpo decided to take his first lesson. This happened by accident when an admirer came backstage after a matinee on Broadway and turned out to be the harpist for the Metropolitan Opera. At Harpo's request the gentleman agreed to teach him for $20 a shot. On arriving at the first lesson Harpo played the sextet from *Lucia*. "The maestro," wrote Harpo, ". . . watched me so closely his nose was practically sticking through the strings, and he kept talking to himself: '*Ah, yes! . . . Ah, so! . . . Extraordinary! . . .*'" The teacher merely shrugged when Harpo suggested that he learn to read notes, and for an hour just watched, analyzed, and learned from Harpo's technique. That, of course, was the end of Harpo's studies. "I was always willing to demonstrate," he conceded, "but damned if I would ever again pay twenty bucks to give any teacher a lesson."

A harpist named Mildred Dilling, whom Harpo met in a music store, introduced him to Bach and Mozart and gave him some pointers. When he couldn't get a particular chord, he'd call her up, no matter what the hour, and she'd put the phone next to her harp and play, until he got the same sound at the other end.

Such virtuosos as Salzedo and Grandjany have been thoroughly mystified by Harpo's unusual technique. However he developed it, whatever long hours of experiment and practice it took, the method was his own and the music he made highly original because of it. "Whenever I touched the strings of the harp," admitted the famous comedian, "I stopped being an actor."

?

How did FDR come to appoint the first woman to a president's Cabinet?

When Roosevelt was assembling his Cabinet in 1932, labor leaders had their own ideas about who should be Secretary of Labor. By the time the President met with them, however, his mind was made up. He had not only rejected their choices, he had selected a woman: Frances Perkins. Eleanor Roosevelt sympathized with her husband when he had to defend the unprecedented appointment before the labor leaders. But when she tried to comfort him, the President shrugged it off, allegedly remarking, "Oh, that's all right. I'd rather have trouble with them for an hour than trouble with you for the rest of my life!"

?

How did they split the atom?

In the early days of atomic physics, the overriding concern was the search for knowledge, pure and simple, unencumbered by the controversial moral and political ramifications that since 1945 have surrounded the atomic bomb. In the forefront of the drive to reveal the secrets of the atom was one of the greatest physicists of all time: Ernest Rutherford (later Lord Rutherford), the exuberant and highly intuitive son of a Scottish immigrant to New Zealand. Rutherford employed simple, direct methods that, by virtue of his farsightedness, led him to ground-breaking results. In 1911, for instance, he discovered the nucleus of the atom simply by intercepting a flow of alpha particles with a thin sheet of gold foil. (These particles are one of three types of radiation shown by the Curies to be emitted from radium. The alpha particles, identical to helium atoms minus their electrons, are electrically charged.) When some of the high-speed particles bounced straight back from the foil, Rutherford reasoned that they were being deflected from hard, massive units within the atoms: the nuclei.

Using alpha particles again in 1919, Rutherford achieved the first transmutation of elements, which led directly to the splitting of the atom. He issued his alpha particles (helium nuclei) down a glass tube filled with nitrogen gas. To his amazement, he detected hydrogen nuclei (later called protons) at the other end, although this element had not originally been present in the tube. His projectiles had actually succeeded in dislodging these protons from the nuclei of the nitrogen atoms. This meant that those highly concentrated nuclei were not indestructible, that physicists might indeed probe their nature and structure. Rutherford noted, furthermore, that the proton that was released had greater energy than the original alpha particle. Here was the first clue to the release of atomic power.

Rutherford theorized that if alpha particles could cause transmutation, other particles might be capable of doing the same, given that they had a comparable amount of energy. He decided to build a machine to accelerate protons and conceived of a positive-ray tube requiring 1 million volts. But in 1930 electrical technology was insufficient to generate and maintain such high voltage. The ingenious physicist therefore suggested to his associates, John D. Cockroft and E.T.S. Wilson, that they attempt the experiment with already existing equipment and relatively low-voltage particles.

These two men worked with lithium, a very light, common element. They built a proton gun, containing a hot filament that emitted electrons. Accelerated by a positive voltage, the electrons shot through a hole to a section of the gun where they struck atoms of hydrogen and caused ionization. A strong negative voltage of 150,000 volts was then applied. This attracted the positively charged protons and repelled the electrons back toward the other end of the gun. Some of the protons passed out of the gun as a beam of 150,000-volt energy particles. Now this beam was directed at the lithium. The protons bombarded the lithium nuclei and combined to raise their atomic mass (i.e., weight) from 7 to 8; the protons also increased lithium's atomic number (the number of electronic units of positive electrical charge in the nucleus) from 3 to 4. This was too much for the lithium nucleus to bear, and it immediately exploded, splitting precisely in two. The result: two helium nuclei of atomic weight 4 and nuclear charge 2—alpha particles. Although the energy input was only 150,000 volts, each of the new man-made alpha particles was emitted with a dazzling energy of 8.5 million volts. Not all the protons succeeded in finding their mark, of course. The lithium nuclei were tiny targets, and the vast majority of protons glanced off to the side, missed entirely, or just lost their energy from hitting other electrons. Nevertheless, the proton gun worked and for the first time the atom was split under man's direct control. The normally self-effacing Cockroft, overwhelmed by the phenomenon, walked through the streets of Cambridge, England, exclaiming to friends and strangers alike, "We've split the atom."

Nineteen thirty-two was also the year of discovery of the third subatomic particle, which Rutherford had predicted some years before. Through exhaustive work night and day over a period of only three weeks, James Chadwick of England proved the existence of the neutron, a particle with mass but no electrical charge. The latter characteristic was highly significant when it came to splitting the atom. Because the neutron is neutral, it can approach the positively charged nucleus without being repelled, and do so even at slow speeds. Positively charged particles, on the other hand, can only get close to the nucleus if fired at tremendous speeds, sufficient to overcome the electric-charge repulsion. It was time to challenge increasingly heavier atoms than the volatile lithium.

Enrico Fermi, a Nobel Prize–winning physicist at the University of Rome, was the first to realize that the neutron could actually be captured and absorbed by the nucleus of an atom, thereby adding to its mass. Fermi found that a fast stream of neutrons resulted from the impact of alpha rays from radium on metallic beryllium. He then slowed the rapidly moving neutrons by having them pass through blocks of carbon. He subsequently made many new artificial atoms by exposing various elements to the intruding neutrons. One of these elements was uranium, and the effects of exposing it were puzzling. The resulting new element produced an abnormal amount of radiation. The nuclei were very heavy—heavier than those of uranium, which is the heaviest natural element on earth. Because they were so heavy, Fermi reasoned, they were unstable; they disintegrated, emitting the radiation he had witnessed. The chemistry of the new atoms did not seem to meet the characteristics of any known element. Thus a number of scientists (Fermi himself was skeptical) believed that new elements, called transuranic, had been discovered. It is terrifying to think how close Fermi and his associates came to blowing themselves up along with a good part of Rome. Had they realized what was in fact before them, the atomic bomb might have been developed as early as 1934, giving Hitler plenty of time to follow suit or, on the other hand, discouraging him from launching World War II at all.

When chemist Otto Hahn of the Kaiser Wilhelm Institute in Berlin repeated Fermi's experiment, his results were the same. But on careful examination of the end products, he found that the common isotope uranium-238, with its 92 protons and 146 neutrons, had become barium, with a much lighter nucleus of 138, comprising 56 protons and 82 neutrons.

It was the persistent, clear thinking of Lise Meitner that penetrated the enigma, which in retrospect should have been obvious from the start. Meitner was Hahn's associate, a Jew who fled Austria in 1938 and arrived in Göteborg, Sweden, where she was joined by her nephew, physicist Otto Frisch. Until then, everyone had assumed that the very heavy nucleus of uranium was stable, that despite its many protons, which repelled each other, there were even more neutrons, which kept the whole unit together. At most, small fragments could be broken off—charged particles like the alpha particles Rutherford had used. Meitner reflected on the analogy of an atom to a drop of water, postulated by the famous Danish physicist Niels Bohr. Just as internal stress can cause a liquid drop to split into two smaller drops, so, Meitner realized, the bombardment of the uranium nucleus with neutrons overloaded the nucleus and caused it to break apart. Unable to incorporate the neutrons, the nucleus became unstable and divided, the two new nuclei still containing electrical charges of the parent. One part was barium, with its 56 electronic units; the other was found to be krypton, with 36 units. These charges repelled each other, as was expected, but the *force* with which they did so proved astounding. Using Einstein's theoretical relationship between mass and energy and precise atomic measurements, Meitner calculated the energy released by the split to be 200 million electron volts, by far the greatest energy source known to man. Uranium fission had been discovered.

Initially, the discovery was not thought to be of much practical use. The splitting of a uranium atom was exceedingly rare, and it was too difficult to sustain a prolonged fire of neutrons. Nevertheless, scientists throughout the world were excited. Niels Bohr and John A. Wheeler of Princeton theorized that when the uranium atom splits, it releases neutrons, which could go on to break up other nuclei. This would cause a chain reaction, an increasingly fast series of fissions. By March 1939 physicists accepted Bohr's theory and his insistence that the rare isotope uranium-235, if collected in large enough quantities, could create a new and massive bomb. The search was on, the wheels of the United States government were set in motion, and the Manhattan Project got under way.

?

How did they name America?

When Christopher Columbus died in 1506, he adamantly believed that the scene of his adventures and exploration was eastern Asia. This conviction prevented his assigning a new name to the vast continents that by rights should bear his name. His error was the first in a chain of quirky occurrences through which a counterfeit letter eventually provided grounds for the naming of America.

Five years after Columbus anchored his three ships off the Bahamas, a quiet, respectable Florentine merchant and astronomer named Amerigo Vespucci made the first of his four voyages to the Western Hemisphere. He landed farther south, on what is now South America, and soon perceived the land to be a continent, and a *new* one. He spent some time among the natives, who astonished him with their nakedness, penchant for taking numerous wives, lack of personal property, and ferocious wars—the reasons for which the navigator could not fathom. His modest and patently unoriginal suggestion for a name for the new territory was Mundus Novus, or New World.

Vespucci liked to describe his experiences and observations in letters to friends back home. Many of these letters apparently went astray and someone, discerning their bestseller potential, sensationalized them and published them as *Four Voyages,* with Vespucci as author. A certain letter published by the Academy of the Vosges in Lorraine in April 1507 was read with particular interest by Professor Martin Waldseemüller, a German cartographer then working at the academy. He promptly published the letter in his new book, *Cosmographiae Introductio,* commenting ". . . Since these parts have been more extensively explored, and another 4th part has been discovered by Americus Vespucius . . . I see no reason why it should not be called Amerigo, after Americus, the discoverer, or indeed America, since both Europe and Asia have a feminine form of name from the names of women."

This book was circulated widely, and Waldseemüller also published a map of his "America," showing what is now Brazil. By the time he discovered his error and the injustice done to Columbus, others had accepted the name and, in fact, made it the official one for all the Western Hemisphere.

In the end, both worthy explorers died in ignorance—Columbus of the true nature and extent of his find, Vespucci of the fact that two enormous continents would immortalize his name.

?

How did they invent cigarettes?

The Aztecs smoked tobacco through a hollow reed or cane tube, and elsewhere in Central and South America the indigenous population rolled shredded tobacco in corn husks and other vegetable materials. The conquistadores who witnessed this returned to Spain with something resembling our present-day cigar. Wealthy Spaniards smoked these cigars and often tossed the butts into the street, to be retrieved by the watchful poor. By shredding old cigar butts and rolling the tobacco in scraps of paper, the beggars of 16th-century Seville created the first cigarettes, which they called *papeletes* or *cigarrillos*.

Several centuries were to pass before the lowly *cigarrillo* moved to wealthier circles. As it did, it also passed across borders to Italy and Portugal, and from there via traders to the Levant and Russia. The French and British became acquainted with them during the Napoleonic Wars and the French gave them the name we use today. Later, during the Crimean War, the French and British discovered fine Turkish tobacco.

The earliest manufacture of cigarettes was by hand, by the smoker himself or by factory workers who rolled them on a table, then pasted and packaged them. In 1880 American inventor James A. Bonsack created the first automatic cigarette-making machine. Tobacco was issued onto a continuous strip of paper, and the cigarette was formed, pasted, closed, and cut with a rotary knife. This method foreshadowed the mass production of cigarettes that would, for better or worse, make cigarettes a common pleasure throughout the world.

?

How did they kill Marie Antoinette?

On October 16, 1793, Marie Antoinette, former Queen of France and Navarre, mounted an open tumbrel led by cart horses, her hands bound behind her. With a constitutional priest beside her, she was driven through the chill autumn morning from the Conciergerie, where she had been held prisoner, over the Pont au Change, toward the rue Saint-Honoré, past the lodgings of the "Incorruptible Patriot" Robespierre, to the place de la Révolution. *"Vive la République!" "À bas les tyrans!"* shouted the hordes who had come out to witness the final degradation of their Queen. Her once luxurious hair, now white, was cropped beneath a bonnet, her sumptuous silks replaced by simple cotton. She calmly climbed the scaffold, then accidentally stepped on the foot of the executioner, Henri Sanson. "*Monsieur,*" said Marie Antoinette, "I beg your pardon. I did not do it on purpose." These were her last words before her head fell beneath the blade of the guillotine, and they are an apt comment on her entire life.

Twenty-three years earlier a young, naïve, and uneducated girl of 15 was brought to France and welcomed by the populace as their future Queen. This was a daughter of the great Empress Maria Theresa of Austria, who, for political reasons, had proposed to Louis XV the marriage of Marie Antoinette and his grandson Louis-Auguste Capet, future King of France. The young couple were married in the splendor of Versailles in 1770, and four years later the Dauphin (Louis-Auguste) was crowned at Reims.

Then began a fairy-tale life for the young Queen, whose extravagances had no limit. Her silks and brocades numbered in the hundreds, a household of 500 ministered to her every need, and each morning a hairdresser drove from Paris in a coach drawn by 6 horses to style her hair. In those days of rococo artifice, women wore their hair combed up on top of their heads in fabulous arrangements, sometimes three feet high, topped with precious objects, which might depict a garden or a hunt. Marie Antoinette had her own palace, the Petit Trianon, where she indulged her lavish tastes and whimsical desires. At "Little Vienna," as she fondly called it, the Queen reigned supreme, posted orders for the servants, kept a certain social set on hand to amuse her, and opened a small theater for further diversion.

From Austria, Maria Theresa watched with increasing apprehension and wrote frequently to her frivolous daughter. "I know only too well how extravagant you are. . . . If you are not careful, you will lose, through these follies, the good will that you had at the beginning. Everybody knows that the King is thrifty, so you will receive all the blame. I hope I do not live long enough to see the disaster which will probably result."

Marie Antoinette paid no attention, and the charming life lapsed into decadence, infested with scandal and intrigue, disturbed by mounting debts. Some of those at the court of Versailles proved opportunistic and disloyal; rumors were spread about the Queen's supposed affairs and misconduct. More importantly, the populace at large, plagued by burdensome taxes and inflation in the years preceding the Revolution, began to see the Queen as symbolic of gross inequality and bureaucratic corruption. She blithely earned the nickname Madame Déficit. The young Queen also meddled in international politics, not always in the best interests of her country. In response to requests from her Austrian relations, Marie Antoinette made concessions for financial and military assistance in both 1778 and 1784—against the wishes of her husband. This and a certain amount of favoritism in her appointment of ministers proved potent ammunition at her trial some years later. The French disliked Marie Antoinette's influence over the King, not only because of what she actually did, but because they could not abide the idea of being ruled by a woman.

If Marie Antoinette had her faults, she cannot be said to have caused all that troubled France. Nevertheless, she and Louis XVI became the obvious scapegoats for decades of oppression, discord, and economic difficulties, exacerbated by the excesses of the regime of Louis XIV. The populace was seething, and the King was ineffectual. On July 14, 1789, the people stormed the Bastille, and the Revolution was under way. They rioted at Versailles on October 5 and 6, abducted the royal family, and brought them to the Tuileries Palace in Paris with the intent that here they might rule more effectively, reform the constitution, and control inflation. Instead the King and Queen attempted an escape, only to be captured again in Varennes. In 1792 mobs stormed the Tuileries in a bloody siege. The National Convention, the second revolutionary assembly, assumed sole authority, brought Louis XVI to trial, and found him guilty of supporting counterrevolution and of killing Frenchmen. He was guillotined in 1793 and "the Widow Capet" confined to the Conciergerie under constant surveillance.

As Marie Antoinette sat in her damp cell without heat, adequate furnishing, books, or even a mirror (which particularly troubled her), the Terror was unleashed on France and revolutionaries Robespierre and Danton reached the height of power. At last the convention agreed that it was "time to root out every trace of monarchy," that the hapless, and by now insignificant, widow should be tried by the Revolutionary Tribunal, whose methods Danton underscored: "Let us be terrible, to spare the people being terrible."

Marie Antoinette was brought to trial on the evening of October 12. Heavily guarded, she was led upstairs to the Palais de Justice, dark except for two candles to light her face. The prosecutors had found virtually no tangible evidence against the Queen and so resorted to popular rumor, a dizzying tirade of truth and lies. The clerk read a lengthy document accusing her of causing the riots of October 5 and 6, 1789, and inciting the King's flight to Varennes, his vetoes of National Assembly decrees, and his deception of the French people. Deputy Public Prosecutor Fouquier-Tinville charged that Marie Antoinette had ordered Swiss Guards to fire on the French who stormed the Tuileries, and that she had obtained their loyalty with numerous bottles of liquor. There was a ludicrous charge that the Queen carried two pistols with her at Versailles in order to shoot the Duc d'Orléans. Time and again the prosecutor referred to the sumptuous banquets and parties that had occurred at Versailles, laying particular weight on the exorbitant sums spent at the Petit Trianon. Furthermore, Marie Antoinette was said to have sent gold to her brother, the Emperor of Austria, for war against the Turks, and to have betrayed French military secrets to foreign powers.

On October 13 the trial was resumed. Forty witnesses were

Stripped of her finery, Marie Antoinette awaits the scaffold. Jacques Louis David sketched the former Queen as she passed his window on October 16, 1793.

called in to testify. Questions and accusations were hurled at Marie Antoinette, to which she responded with remarkable calm and resourcefulness.

"Do you regard a King essential to the welfare of the nation?"

"An individual cannot judge of that."

"Do you regret your son's loss of the crown?"

"I shall regret nothing for him when France shall have been made happy. . . ."

"During the revolution you have not ceased to work against liberty, to intrigue with foreign powers."

"Since the beginning of the revolution I have denied myself all foreign correspondence, and I have never had a decisive voice in the internal affairs of government." Here Marie Antoinette blatantly lied, and all who were present knew it.

Finally, just as Anne Boleyn was charged with incest with her brother when Henry VIII wanted to get rid of her, Marie Antoinette was charged with having such a relationship with her eight-year-old son. Assistant Prosecutor Hébert produced a signed confession that he had extorted from the frightened Dauphin. To support this preposterous claim, Hébert ascribed to the Queen an even more hideous motive: that she wished to injure her son, weaken his powers, and thus continue to dominate him even on the throne.

Some of the charges of the Revolutionary Tribunal had some truth, while others were blatantly ridiculous, but the outcome was predictable from the start. Marie Antoinette was pronounced guilty of instigating plots to assist foreign powers and outside enemies financially, thus strengthening their armies, and to start a civil war within the Republic of France. She was condemned to death at 4 A.M. on October 16, 1793, leaving her only a few hours in her cell before facing the angry mob that had once applauded their Queen.

?

How did they know the size of the earth over 1,700 years before anyone sailed around it?

In the third century B.C., Eratosthenes needed no more than a gnomon, or shadow stick, to measure the whole earth, and his answer missed our contemporary, sophisticated measurements by only 5 percent.

On the summer solstice, Eratosthenes stood at Alexandria with his gnomon. When the sun reached its zenith over Syene, a city 5,000 stades, or 527 miles, due south of Alexandria, he measured the length of the shadow cast by his stick. At Alexandria the angle of the sun was 7°12′ away from its zenith. From this observation and knowledge of the distance to Syene, he calculated the earth's circumference as 250,000 stades, or 26,350 miles. The actual polar circumference, according to scientists today, is 24,859 miles.

?

How did they determine the membership of the U.S. Senate?

The years preceding the enactment of the Constitution saw the need for a stronger national government. The Congress that had been active since 1774 was weak, and while it technically had the right to coin money, regulate Indian affairs, declare war and peace, and appropriate funds, it had little power to carry out these responsibilities. Each state had one vote in Congress, which was determined by a majority of the delegates from that state (there had to be at least two), who were appointed for one-year terms by the state legislatures. Representation from at least seven states, and in some instances nine, was required to accomplish anything, and this proved a problem. Representatives were constantly ill or absent for one reason or another; those who did show up sometimes waited weeks for others to appear. The brevity of their terms contributed further to the instability of this governing body. National unrest was the result. Farmers in the West rose up against burdensome state taxes and restrictions. Spain refused Americans the right to navigate the Mississippi and, hungry for territory, wooed the troubled farmers in the areas that are now Kentucky and Tennessee. A secessionist movement threatened North Carolina. States were bickering over their charters and boundaries, for some were designated as stretching all the way to the Pacific, while others were not. The economy was increasingly unstable, the paper money printed by Congress almost worthless. Congress, furthermore, failed to make satisfactory economic treaties with England.

Although proposals for a more binding central government were made as early as 1776, little was done because many feared the crushing authority of a monarchy, such as they had recently freed themselves from. State legislatures, which were daily assuming more power, naturally encouraged this sentiment. By the latter part of the 1780s, which brought Shays' Rebellion and several instances of national humiliation, many members of Congress, political leaders, and writers were realizing the need to think, as Alexander Hamilton put it, "continentally." A decision was made by all states except Rhode Island to revise the Articles of Confederation, the existing framework of the government, and in 1787, 55 delegates—including Benjamin Franklin, George Washington, James Madison, and Hamilton—arrived in Philadelphia to do so.

Edmund Randolph of Virginia presented the plan of his brilliant colleague Madison, then only 36. This plan attempted to bypass the state legislatures and form a government directly responsive to the people, with representation proportionate to the population. (At the time each state, no matter what its size, had one vote in Congress.) Most significantly, this new government, unlike any other, would have a built-in system of checks and balances so that no single branch could amass all the power. After two weeks of discussion and revision, most of the convention agreed that the legislative branch should comprise two houses: members of the upper would be appointed by the state legislatures, members of the lower elected by the people. The number of representatives in each would be proportionate to the population.

This plan might be in effect today had it not been for William Paterson (remembered perhaps by residents of the city in New Jersey named for him), who spoke up on behalf of the disgruntled small states for equal representation in at least one house. Only when delegates from these small states threatened to quit the convention altogether did the other delegates agree that the upper house should allow each state an equal vote. In return for this concession, the large-state delegates determined that all appropriation of funds should be empowered to the lower house, and taxation would correspond to the population, which would be counted by a census every ten years. Each state legislature would appoint two senators to serve in the upper house for six years; every two years the people would elect one representative for every 30,000 persons to the lower house; and the two houses would be equal in power. Anticipating some opposition to the new plan, particularly by existing state legislatures, the convention proposed that the people in each state elect a convention to decide on the Constitution, which would become effective as soon as nine states had accepted it. After almost a year of national debate in which the proponents of the new plan acclaimed it in speeches and writing, the Constitution was ratified. Thus the Congress succeeded in dissolving itself and opening the way for a new government body, as determined in the first national elections in January 1789. It was not until 1912 and the passage of the 17th Amendment to the Constitution that senators came to be elected by popular vote instead of legislative ballot.

?

How did Hannibal cross the Alps?

Hannibal, born in 247 B.C., grew up in Carthage during the First Punic War between the great powers of Rome and Carthage, rivals since the days long before when Dido of Libya had died of love for Rome's Aeneas. At the tender age of nine, Hannibal vowed a sacred oath before his father, the renowned general Hamilcar, to hate Rome forever, and he fueled this hatred throughout his life, developing a passion for war as he abstained from wine and women. The Second Punic War was launched when Hannibal, as general of Carthaginian troops in Spain, seized the Roman-allied town of Saguntum in a bloody massacre and brazenly undertook a march across the Alps to meet the Romans on Italian soil.

Hannibal's crossing these snowy peaks with pack animals and thousands of troops is a feat that has amazed succeeding generations for centuries, an undertaking that attested to the general's staunch will and ingenuity. Hannibal was greatly admired for his daring and readiness to forgo comforts, and Livy described him as an inspiration to his men. Throughout the march he encouraged and reassured his troops, a scurrilous assortment of mercenaries, who certainly did not remain with him out of patriotic sentiment. Hannibal's steadfast leadership through barbarian attacks and perilous mountain passes was a key ingredient in his success.

Neither Hannibal nor any other African recorded his arduous trek. We rely on two Roman sources, Polybius and Livy, whose writings provide the basis for the following account. Further credit is due Sir Gavin de Beer, who spent 40 years of scholarly research tracing Hannibal's footsteps.

Hannibal left Cartagena on the southern coast of Spain in May of 218 B.C., with 90,000 foot soldiers, 12,000 horsemen, and many elephants. A battle with stubborn tribes in Catalonia reduced his numbers somewhat, and he was forced to leave 10,000 men and 1,000 horses to control the passes over the western Pyrenees, between Spain and Gaul. Three thousand more soldiers deserted him at the foot of the Pyrenees when the alarming extent of his ambitions became known. The general thereupon got rid of 7,000 more, considered unreliable, and proceeded over the eastern Pyrenees, perhaps via the 360-meter-high Col de Banyuls, with 50,000 infantry, 9,000 cavalry, and 37 elephants. Nearly all these elephants were of the African type, smaller than the Indian elephant and more agile in negotiating steep mountains and rocky terrain. Elephants had been highly prized in war for over a century, but Hannibal not only intended to use them against cavalries whose horses had not been trained to meet them, he knew the novelty and size of the elephants would terrify ill-willed tribes along the way.

The army marched through southern Gaul somewhere near the coast, traversing marshes and lagoons, until it reached the Rhone River, then headed north for four days to Fourques, where the river is somewhat calmer and shallower. The physical problem of getting bag and baggage as well as elephants across the river was exacerbated by the hostile Volcae, a Gallic tribe crowding on the eastern bank, just waiting to pick off the Carthaginians when they were most vulnerable, in midstream. But the crafty Hannibal sent a unit of Spaniards 22 miles upstream. Using rafts made from tree trunks or merely the aid of their shields, the men floated and swam across. Once on the western shore they headed south, coming up on the Volcae from behind. When Hannibal received a smoke signal from his confederates at dawn, he began to cross the ford, while the dumbfounded Volcae were easily routed from behind. Hannibal's men, meanwhile, had bribed the tribes on the west bank for the use of some boats, and had built rafts and hollowed out tree trunks as well. To get the elephants across, they constructed piers that jutted 200 feet into the river and were secured by ropes tied to trees upstream. Large rafts with floats were than attached to the piers and covered with earth to fool the elephants, which wouldn't take to the idea of leaving dry land. The females were led on first, and the males happily followed. But then some of the elephants panicked, stampeded, and plunged into the water. Fortunately the river was just shallow enough so that, with their trunks high over their heads, the elephants managed to wade to shore. Horses swam across, led behind rafts, and the men floated or swam.

Before heading east Hannibal marched farther upriver to "the Island," a fertile triangle of land bordered by the Rhone and

The approximate route of Hannibal's rugged march in 218 B.C. from Spain over the Alps to Italy.

ALPS
Isère
CISALPINE
GAUL
Insubres
Cremona
Po
Clastidium
Placentia
Mutina
Durance
LIGURIA
Massalia
Ariminum
ADRIATIC SEA
Faesolae
ETRURIA
Arretium
L. TRASIMENE
Telamon
CORSICA
Alalia
Caere
Pyrgi
Rome
Cannae
SARDINIA
CAMPANIA
TYRRHENIAN
SEA
Crotona
LIPARI IS.
Mt Eryx
Panormus
Drepanum
AEGADIAN IS.
Lilybaeum
Messana
SICILY
Carthage
Agrigentum
CAPE
ECNOMO
Bagrades (Medjerda)
Tunis
A
Zama
Hadrumetum (Sousse)
MALTA
Leptis
(Lebda)
CERCINA IS.

Aygues rivers and the Baronnies Mountains. There Hannibal diplomatically settled a kingship dispute between two brothers. The one who thereby came to power gratefully supplied Hannibal with wheat and clothing and provided an escort to fend off the hostile Allobroges tribe, which occupied vast territories to the north.

This escort accompanied Hannibal north, then east along the Drome River to the ascent to the Alps and the Col de Grimone. On their departure, Hannibal became aware that the Allobroges were nearby (this prior knowledge probably saved him from being defeated altogether) and sent several Gauls to discover the barbarians' intentions. The spies learned that the Allobroges guarded the mountain pass by day, but retired to their town at night. Hannibal thereupon moved his camp openly to a point dangerously near the Allobroges and proceeded at night with a group of his best men through the gorge, occupying the exact position held by the barbarians by day. The Allobroges were stunned and might have drawn back had they not then seen the remainder of the army slowly plodding through the gorge, ripe for the taking. They attacked from above, wreaking havoc, causing many to fall from the narrow path along the cliffs. Hannibal rushed back and managed, after heavy losses, to beat off the powerful tribe.

The battered army marched peacefully through the wide Bassin de Gap to the Durance River, where it crossed over and followed a watershed pass leading out of the valley, probably along the Guil River through the Queyras and over the Col de Traversette. But a second battle with barbarians hampered its progress. On the seventh day since his ascent into the Alps, Hannibal was greeted by tribes bearing olive branches and wreaths. Suspicious, Hannibal pretended to befriend them and received hostages who acted as guides and led the unwieldy army through an extremely narrow pass, soaring cliffs to one side, a sheer drop on the other. As Hannibal's men and animals navigated the treacherous path, the alpine tribes sneaked up from behind. The general, anticipating as much, had placed heavy infantry in the rear, which fought them off. The situation became desperate when the marauding tribes hurled down boulders from above, causing thousands to fall to their death and splitting Hannibal's column in two. Suffering heavy losses, the weary troops were forced to continue through the gorge all night.

Hannibal marched over the main watershed pass, towering some 9,680 feet high, and camped in the snowy peaks for two days. There he tried to rally his dismayed army with a view of Italy's Po Plain stretching in the distance far below. Barbarians no longer threatened them but the precipitous and slippery descent proved just as fatal. Many lost their balance and perished; scores of animals whose legs penetrated the new soft snow got stuck in the hard, frozen snow underlying it. A famous detail from Livy describes the removal of a gigantic boulder that stood in the way. The men cut trees and covered the rock with logs, to which they set fire. Then they poured vinegar (sour wine) over the rock, which made it more likely to crumble, and attacked it with pick-axes. The sight must have been a dramatic one, the massive rock burning in the remote, icy peaks as men and animals huddled in the cold.

When, after 15 days in the Alps, Hannibal reached the plains below, he had 12,000 African infantry, 8,000 Spaniards, 6,000 cavalry, and all 37 elephants. The hardy survivors had covered 1,500 kilometers from Cartagena in five months—only to rest three days and then undertake a futile war with a country capable of massing three quarters of a million troops against them.

?

How did they discover radium?

The story of the search for radium is a romantic and stirring one. Behind it is a woman who was passionately curious, daring in her convictions, and determined to work in an age hardly encouraging to professional aspirations among those of her sex. From a dilapidated shed, described by one German chemist at the time as a "cross between a stable and a potato-cellar," came a discovery that would throw light on the structure of the atom, open new doors in medicine, and save lives in future generations.

Marie Sklodowska came to Paris and the Sorbonne in 1891 as a reticent Polish woman of 24. Taking a solitary room in the Latin Quarter, she began her studies in mathematics and physics. By 1897 she had two university degrees and a fellowship, as well as a husband and a newborn daughter. In the physicist Pierre Curie, Marie had found both an adviser and a lover, someone as serious as she, who shared her interests and became drawn into her quest.

In that year Marie was casting about for an appropriate subject of study for her doctorate. The scientific world was in an uproar over Wilhelm Röntgen's discovery of X rays in 1895, and this was an obvious field. But Marie looked further, and an accidental discovery by Henri Becquerel sparked her interest. In 1896 Becquerel was studying uranium salts. He had wrapped some photographic plates in a black cloth and covered them with a sheet of aluminum, followed by crystals of a uranium compound. He planned to study the effects of the sun on the crystals, but since it was cloudy that day and on the two succeeding days he left the plates in a drawer. On removing them he was stunned to see that the plates were fogged where the crystals had covered them—that is, the uranium had produced an impression on the plate similar to that which light would make. Becquerel was the first to see evidence of the spontaneous emission of rays, what Marie Curie would later name radioactivity.

The intriguing question, of course, was where did this energy come from? What was the source of energy that allowed the rays to penetrate not only paper, but metal? An ambitious Marie Curie undertook to find out. Through her husband and the School of Physics and Chemistry, Marie obtained a small room in which to work, a glassed-in storeroom housing lumber and machines, on the ground floor of the school. It was damp, ill equipped, even lacking adequate electricity. But Marie set to work, first to determine the power of ionization of uranium—that is, its capacity to make the air conduct electricity. She used an electrometer called a piezoelectric quartz, which Pierre and Jacques Curie had devised some years before. Realizing that a crystal will be slightly deformed by an electric charge applied to it, the brothers developed an instrument that amplified this deformation; thus they could measure small electric currents. Marie's experiment using the piezoelectric quartz was quite simple. She placed the substance to be tested in a chamber of ionization consisting of a lower plate, which was charged, and another plate opposite to it, which was attached to the electrometer. She tried to make an electric current pass through the air between the plates, and if this occurred, the electrometer would detect it. Within several weeks she found that uranium's ionizing capacity was an innate characteristic, which neither resulted from a chemical reaction nor depended on external circumstances such as light, moisture, or temperature. She had shown, in fact, that the activity of uranium was an atomic property of that element, a ground-breaking find for 20th-century physics. Mme. Curie also found that the intensity of the rays was proportional to the quantity of uranium present in the compounds she examined.

If uranium had such power, why not other materials? Was it purely chance that uranium's unusual characteristic had been revealed? Marie energetically gathered materials from the School of Physics with the intent to study all known elements, whether in a pure or compound state. Before long she discovered that thorium, too, emitted rays, of an intensity comparable to those of uranium.

Fortunately, Mme. Curie did not stop with these discoveries. She wanted to test still other materials and obtained various mineral samples from the School of Physics. As might be expected, those containing thorium or uranium were radioactive; the others were not. But what came as a bewildering surprise was that the intensity of emissions by the active minerals—specifically pitchblende—was four times that of a pure oxide of uranium. Marie was sure her experiment was erroneous and repeated it scores of times. *Something* was far more radioactive than uranium or thorium, yet Marie had already tested all known elements.

On April 12, 1898, Mme. Curie's former professor, Gabriel Lippmann, made a presentation to the Académie des Sciences on her behalf. In it was expressed a "belief that these minerals may contain an element which is much more active than uranium." Curie's hypothesis, then, was the existence of a *new element*.

At this point the young scientist was overwhelmed by excitement and passion to prove her theory. Pierre, too, was intrigued enough to abandon his study of crystals to join her. Initial experiments with small amounts of pitchblende were begun. Marie and Pierre used a new method of chemical research based on radioactivity; they separated the substance into various products by traditional means of chemical analysis and then measured the radioactivity in each. As the breakdown continued, the radioactive element became increasingly concentrated. It was possible to separate it from all except one element: bismuth. By June 1898 Marie had a sample of impure bismuth sulphide that was 150 times more radioactive than uranium. Pierre placed this substance in a tube and heated it. The substance itself remained in the hotter part of the tube while a thin black powder collected apart from it. This potent powder showed an activity 330 times greater than uranium.

In the *Proceedings of the Academy* of July 1898, the Curies wrote, "If the existence of this new metal is confirmed we propose to call it *polonium,* from the name of the original country of one of us."

November brought a startling development. Upon removing bismuth and polonium from the radioactive material in pitchblende, they obtained a liquid containing barium, which was still highly radioactive. After dissolving and reprecipitating it numerous times, they had a residual substance 900 times as radioactive as uranium, yet barium itself was known *not* to be radioactive. The scientists strongly suspected the presence of a new element. They now turned to Eugène Demarçay, who operated a spectroscope, which was the means to confirm their suspicions. Demarçay dissolved a tiny sample of the Curies' barium compound in acidified water and painted it onto electrodes. He passed a spark along the electrodes and was able to photograph the spark spectrum of the substance. He found a spectral line (or linear image corresponding to a component of the radiation emitted by the substance) that did not result from any known element. As the solid was purified, furthermore, the spectral line grew more intense. On December 26, 1898, the Curies announced a new chemical element in pitchblende: radium, whose radioactivity was thought to be "enormous."

This announcement did not mark the end of Marie Curie's work. Rather, she now set herself the Herculean task of isolating the element so that chemists might *see* once and for all proof of her beloved new element. Tremendous obstacles stood in the way: an inadequate workplace, poor equipment, and a paltry budget. The government was not inclined to open its purse to the cause of science. The two were forced to make do.

It was by now apparent that the amount of radioactive element in pitchblende was so slight (actually only 0.000001 percent) that huge quantities of the mineral would be required to obtain even a negligible amount. Pitchblende ore with its valuable uranium was, furthermore, extremely expensive. The Curies learned that it originated chiefly in the St. Joachimsthal mine in Bohemia, then under the Austrian Empire. The uranium salts were extracted for the manufacture of glass, but what about the residue? The ingenious Mme. Curie offered to buy the cumbersome and useless material and, aided by Professor Eduard Suess in Vienna, her request was fulfilled, free of all costs except transportation.

Early in 1899 a huge horse-drawn wagon loaded with sacks of pitchblende, still mixed with pine needles and dirt, arrived at the School of Physics. The old studio could no longer accommodate them, and the new laboratory the Curies moved to was, if possible, worse. Marie and Pierre spent virtually four years in a tumbledown shed across the courtyard from their former studio. The skylight roof not only leaked, but had no hoods to carry off the poisonous gases from their work. When weather permitted they worked outdoors, but in rain and snow they were forced indoors, opening all the windows whatever the temperature. The shed did not even have a floor, and its furnishings consisted only of a few wooden tables, a blackboard, and an ineffectual cast-iron stove. They suffered freezing temperatures in winter, stifling heat in summer, on top of mysterious ailments they had been trying to ignore in recent years: fatigue, body pains, sore fingers—all symptoms of radiation sickness, of which the two were unwitting victims.

While Pierre studied the properties of radium, Marie became a one-woman factory. Working with batches as large as 20 kilograms, she ground the pitchblende, dissolved it, and filtered it innumerable times. She heaved the materials and stood in the courtyard day after day, month after month, stirring cauldrons of boiling liquids with a huge iron bar. She extracted barium in a chloride state, which she submitted to fractional distillation in porcelain bowls, set up along her wooden tables. Her frustrations were multiplied as airborne impurities—iron and coal dust—inevitably contaminated her rare material.

"We lived in a preoccupation as complete as that of a dream," wrote Mme. Curie. "Sometimes we returned in the evening after dinner for another survey of our domain. Our precious products for which we had no shelter, were arranged on tables and boards; from all sides we could see their slightly luminous silhouettes, and these gleamings, which seemed suspended in the darkness, stirred us with ever new emotion and enchantment."

The moments of enchantment must have been brief, however, compared with her years of tedious labor. But Mme. Curie's singular energy and sheer will prevailed: in March 1902 she had a decigram of pure radium. Demarçay's electroscope had reacted violently to the substance, from which barium at last had been removed. Mme. Curie attributed an atomic weight of 225.93 to her new substance, whose reality and individuality were now confirmed beyond doubt.

Bibliography

HOW DID THEY SPEND $40 MILLION MAKING *HEAVEN'S GATE*?

Ansen, David, with Martin Kasindorf. "*Heaven* Turns into Hell." *Newsweek*. Dec. 1, 1980.

Denby, David. "The Second Time Around." *New York*. May 11, 1981.

Kael, Pauline. "Poses." "The Current Cinema." *The New Yorker*. Dec. 22, 1980.

McGee, Rex. "Michael Cimino's Way West." *American Film*. Vol. 6, no. 1 (Oct. 1980).

New York Times, The. Nov. 20 and 21, 1980.

Stanbrook, Alan. "Hollywood's Crashing Epics." *Sight and Sound*. Vol. 50, no. 2 (spring 1981).

HOW DID THEY DISCOVER PENICILLIN?

Encyclopaedia Britannica.

Ratcliff, J.D. *Yellow Magic: The Story of Penicillin*. New York: Random House, 1945.

Sokoloff, Boris. *The Miracle Drugs*. New York: Ziff-Davis Publishers, 1949.

HOW DID THEY DECIDE TO MAKE WASHINGTON THE CAPITAL OF THE UNITED STATES?

Encyclopaedia Britannica.

Moore, Charles. *Washington Past and Present*. New York: Century Co., 1929.

Wilson, Rufus Rockwell. *Washington: The Capital City*. Philadelphia: J. B. Lippincott & Co., 1901.

HOW DID THEY FIND THE DEAD SEA SCROLLS?

Wilson, Edmund. *The Dead Sea Scrolls, 1947–1969*. New York: Oxford University Press, 1969.

HOW DID RUDOLPH VALENTINO SPEND HIS WEDDING NIGHTS?

Wallace, Irving, Amy Wallace, David Wallechinsky, and Sylvia Wallace. *The Intimate Sex Lives of Famous People*. New York: Dell Publishing Co., 1981.

HOW DID THEY SHOOT UP FAYE DUNAWAY IN *BONNIE AND CLYDE*?

Culhane, John. *Special Effects in the Movies: How They Do It*. New York: Ballantine Books, 1981.

HOW DID THEY KNOW THAT THE DODO WAS EXTINCT?

Alison, Jolly. "Island of the Dodo Is Down to Its Last Few Native Species." *Smithsonian*. June 1982.

Greenway, James C., Jr. *Extinct and Vanishing Birds of the World*. New York: American Committee for International Wildlife Protection, 1958.

HOW DID THEY PUT OUT THE SAN FRANCISCO FIRE OF 1906?

New York Times, The. April 19–21, 1906.

Smith, Dan. *Dan Smith's History of Firefighting in America*. New York: Dial Press, 1978.

HOW DID EVERYBODY START TWISTING?

Carpozi, George, Jr. *Let's Twist*. New York: Pyramid Books, 1962.

Goldman, Albert. *Disco*. New York: Hawthorn Books, 1977.

Life. Nov. 24, 1961.

New York Times, The. May 9, 1966.

New York Times Magazine, The. Dec. 3, 1961.

HOW DID THEY GET THE STONES IN PLACE AT STONEHENGE?

Balfour, Michael. *Stonehenge*. New York: Charles Scribner's Sons, 1979.

Hawkins, Gerald S. *Beyond Stonehenge*. New York: Harper & Row, 1973.

Hawkins, Gerald S., with John B. White. *Stonehenge Decoded*. New York: Doubleday & Co., 1965.

Ivimy, John. *The Sphinx and the Megaliths*. New York: Harper & Row, 1975.

HOW DID MARIE ANTOINETTE "LET THEM EAT . . ." CROISSANTS?

Anthony, Katharine. *Marie Antoinette*. New York: Alfred A. Knopf, 1933.

Trager, James. *The Enriched, Fortified, Concentrated, Country-Fresh, Lip-Smacking, Finger-Licking, International, Unexpurgated Foodbook*. New York: Grossman Publishers, 1970.

HOW DID THEY BLOCK ROOSEVELT'S PLAN TO PACK THE SUPREME COURT?

Blum, John M., Bruce Catton, Edmund S. Morgan, Arthur M. Schlesinger, Jr., Kenneth M. Stampp, and C. Vann Woodward. *The National Experience*. New York: Harcourt, Brace & World, 1968.

HOW DID THE INDIANS DECIDE THAT COWS WERE SACRED?

Auboyer, Jeannine. *Daily Life in Ancient India: From Approximately 200 B.C.–700 A.D.* New York: Macmillan Co., 1965.

Basham, A.L. *The Wonder That Was India*. New York: Grove Press, 1954.

Lamb, Beatrice Pitney. *India: A World in Transition*. New York: Frederick A. Praeger, 1963.

Mayo, Katherine. *Mother India*. New York: Harcourt, Brace & Co., 1927.

Tannahill, Reay. *Food in History*. New York: Stein & Day, 1973.

HOW DID THEY START THE *GUINNESS BOOK OF WORLD RECORDS*?

Wallechinsky, David, and Irving Wallace. *The People's Almanac #3*. New York: William Morrow & Co., 1981.

HOW DID THEY DISCOVER CHAMPAGNE?

Lichine, Alexis, with William Fifield. *Encyclopedia of Wines and Spirits*. New York: Alfred A. Knopf, 1967.

Ray, Georges. *The French Wines*. New York: Walker & Co., 1965.

Waugh, Alec. *In Praise of Wine and Certain Noble Spirits*. New York: William Sloane Associates, 1959.

HOW DID JAMES JOYCE SUPPORT HIMSELF WHILE WRITING *ULYSSES*?

Ellmann, Richard. *James Joyce*. New York: Oxford University Press, 1959.

Ellmann, Richard, ed. *Selected Letters of James Joyce*. New York: Viking Press, 1975.

HOW DID THE FIRST EUROPEAN EXPLORERS CROSS NORTH AMERICA AND REACH THE PACIFIC?

Blum, John M., Bruce Catton, Edmund S. Morgan, Arthur M. Schlesinger, Jr., Kenneth M. Stampp, and C. Vann Woodward. *The National Experience*. New York: Harcourt, Brace & World, 1968.

Clark, William R. *Explorers of the World*. New York: Natural History Press, 1964.

Mirsky, Jeannette. *The Westward Crossings*. New York: Alfred A. Knopf, 1946.

HOW DID THEY DECIDE THAT BLUE WAS FOR BOYS, PINK FOR GIRLS?

Brasch, R. *How Did It Begin?* New York: David McKay, 1965.

Wallechinsky, David, and Irving Wallace. *The People's Almanac #3*. New York: William Morrow & Co., 1981.

HOW DID DICK FOSBURY CHANGE THE TECHNIQUES OF HIGH JUMPING?

Blount, Roy, Jr. "Being Backward Gets Results." *Sports Illustrated*. Feb. 10, 1969.
Christian Science Monitor, The. May 4, 1968.
New York Times, The. Oct. 22 and 23, 1968.
Sports News. July 6, 1968.

HOW DID THEY GET THE IDEA THAT SHAKESPEARE DID NOT WRITE SHAKESPEARE'S PLAYS?

Friedman, William F. and Elizebeth S. *The Shakespearean Ciphers Examined*. Cambridge, England: Cambridge University Press, 1957.
Gibson, H.N. *The Shakespeare Claimants*. New York: Barnes & Noble, 1962.

HOW DID CARDINAL RICHELIEU, PRIME MINISTER OF FRANCE UNDER LOUIS XIII, GET HIS EXERCISE?

Wallechinsky, David, and Irving Wallace. *The People's Almanac #3*. New York: William Morrow & Co., 1981.

HOW DID THEY DESIGN THE FIRST CAR?

Calder, Ritchie. *Leonardo and the Age of the Eye*. New York: Simon & Schuster, 1970.

HOW DID THEY START THE BAADER-MEINHOF GANG—AND HOW DID THE POLICE CATCH THEM?

Christian Science Monitor, The. June 12, 1972.
New York Times, The. June 2, 8, and 12, 1972; Nov. 27, 1974.
New York Times Magazine, The. May 11, 1975.
Time. Feb. 7, 1972; June 12, 1972; Dec. 9, 1974; May 24, 1976; Sept. 19, 1977; Oct. 24 and 31, 1977; Nov. 7. 1977.

HOW DID THEY LAY THE FIRST TRANSATLANTIC CABLE?

Clarke, Arthur C. *Voice Across the Sea*. New York: Harper Brothers, 1958.
Encyclopaedia Britannica.

HOW DID THEY KNOW THE EARTH IS ROUND?

Mitton, Simon, ed. *The Cambridge Encyclopedia of Astronomy*. New York: Crown Publishers, 1977.

BUT HOW DID THEY KNOW THE EARTH IS FLAT?

Dreyer, J.L.E. *A History of Astronomy from Thales to Kepler*. 2nd ed., rev. W. H. Stahl. New York: Dover Publications, 1953.

HOW DID THEY BREAK THE JAPANESE SECRET CODE IN WORLD WAR II?

Clark, Ronald W. *The Man Who Broke Purple*. Boston: Little, Brown & Co., 1977.
Encyclopedia Americana.
Kahn, David. *The Codebreakers*. New York: Macmillan Co., 1967.

HOW DID THEY PICK THE FOUR HUNDRED?

Churchill, Allen. *The Upper Crust*. Englewood Cliffs, N.J.: Prentice-Hall, 1970.
Knickerbocker, Jacob (pseud.). *Then and Now*. Boston: Bruce Humphries, 1939.
Morris, Lloyd. *Incredible New York: High and Low Life of the Last Hundred Years*. New York: Random House, 1951.
Simon, Kate. *Fifth Avenue: A Very Social History*. New York: Harcourt Brace Jovanovich, 1978.
Van Rensselaer, Mrs. John Kine, with Frederic van de Water. *The Social Ladder*. New York: Henry Holt & Co., 1924.

HOW DID THEY DISCOVER THE ATOM IS MOSTLY EMPTY SPACE?

Jastrow, Robert. *Red Giants and White Dwarfs: The Evolution of Stars, Planets, and Life*. New York: Harper & Row, 1967.

HOW DID BALLET DANCERS START DANCING ON THEIR TOES?

Clarke, Mary, and Clement Crisp. *The History of Dance*. New York: Crown Publishers, 1981.

Conyn, Cornelius. *Three Centuries of Ballet*. New York: Elsevier Press, 1953.

HOW DID THEY START THE BREAKFAST CEREAL INDUSTRY?

Trager, James. *The Enriched, Fortified, Concentrated, Country-Fresh, Lip-Smacking, Finger-Licking, International, Unexpurgated Food Book*. New York: Grossman Publishers, 1970.

HOW DID THEY PERFORM THE FIRST CESAREAN DELIVERY?

Cianfrani, Theodore. *A Short History of Obstetrics and Gynecology*. Springfield, Ill.: Charles C. Thomas Publisher, 1960.

Donovan, Bonnie. *The Cesarian Birth Experience*. Boston: Beacon Press, 1977.

Encyclopaedia Britannica.

Time, June 14, 1982.

HOW DID THE BLUE BLAZER BECOME A CLASSIC?

Esquire. June 1982.

HOW DID THEY DECIDE WHICH HORSES WERE THOROUGHBREDS?

Encyclopaedia Britannica.

Encyclopedia Americana.

Hope, C.E.G., and G. N. Jackson, eds. *The Encyclopedia of the Horse*. New York: Viking Press, 1973.

Time. Aug. 2, 1982.

HOW DID HOUDINI ESCAPE FROM A PACKING CASE UNDERWATER?

Cannell, F.C. *The Secrets of Houdini*. New York: Dover Publications, 1973.

HOW DID THEY DISCOVER NEANDERTHAL MAN?

Leakey, Richard E. *The Making of Mankind*. New York: E. P. Dutton, 1981.

HOW DID THEY BUILD THE EIFFEL TOWER?

Harriss, Joseph. *The Tallest Tower: Eiffel and the Belle Epoque*. Boston: Houghton Mifflin Co., 1975.

Poirier, René. *The Fifteen Wonders of the World*. New York: Random House, 1961.

HOW DID THEY DECIDE THAT DECEMBER 25 WAS CHRISTMAS?

Tuleja, Tad. *Fabulous Fallacies*. New York: Harmony Books, 1982.

HOW DID HUMPHREY BOGART CURE A HANGOVER?

Bacon, James. *Hollywood Is a Four-Letter Town*. Chicago: Henry Regnery Co., 1976.

HOW DID THE FBI DEVISE THE "TEN MOST WANTED" LIST?

"Ten Most Wanted Fugitives" Program. Washington, D.C.: U.S. Department of Justice, Federal Bureau of Investigation, 1982.

HOW DID THEY DEVELOP THE SALK VACCINE?

Calder, Ritchie. "Man's Struggle Against Poliomyelitis." *World Health*. Vol. 14 (1961).

Carter, Richard. *Breakthrough: The Saga of Jonas Salk*. New York: Trident Press, 1966.

Encyclopaedia Britannica.
Marks, Geoffrey, and William K. Beatty. *The Story of Medicine in America*. New York: Charles Scribner's Sons, 1973.

HOW DID THEY NAME THE EDSEL?

Brooks, John. "The Edsel: I—The E-Car Has Faith in You, Son." *The New Yorker*. Nov. 26, 1960.
Nevins, Allan, and Frank Ernest Hill. *Ford: Decline and Rebirth, 1933–1962*. New York: Charles Scribner's Sons, 1963.

HOW DID THEY CHOOSE VIVIEN LEIGH TO PLAY SCARLETT O'HARA IN *GONE WITH THE WIND*?

Edwards, Ann. *Vivien Leigh*. New York: Simon & Schuster, 1977.
Robertson, Patrick. *Movie Facts and Feats: A Guinness Record Book*. New York: Sterling Publishing Co., 1980.

HOW DID THEY BUILD CENTRAL PARK?

Barlow, Elizabeth. *The Central Park Book*. New York: Central Park Task Force, 1977.
Reed, Henry Hope, and Sophia Duckworth. *Central Park: A History and a Guide*. New York: Clarkson N. Potter, 1967.

HOW DID PEARY REACH THE NORTH POLE?

Clark, William R. *Explorers of the World*. New York: Natural History Press, 1964.
Peary, Robert E. *The North Pole*. Philadelphia: J. B. Lippincott Co. and the Estate of Joseph Peary, 1937.

HOW DID THEY NAME THE LOLLIPOP?

Wallechinsky, David, and Irving Wallace. *The People's Almanac #1*. New York: Doubleday & Co., 1975.

HOW DID THEY BUILD THE MOTHER SHIP IN *CLOSE ENCOUNTERS*?

Culhane, John. *Special Effects in the Movies*. New York: Ballantine Books, 1981.

HOW DID THEY DISCOVER DNA?

Ebert, James D., Ariel G. Loewy, Howard A. Schneiderman, and Richard S. Miller. *Biology*. New York: Holt, Rinehart & Winston, 1973.
Jastrow, Robert. *Red Giants and White Dwarfs: The Evolution of Stars, Planets, and Life*. New York: Harper & Row, 1967.
Portugal, Franklin H., and Jack S. Cohen. *A Century of DNA: A History of the Discovery of the Structure and Function of the Genetic Substance*. Cambridge, Mass.: MIT Press, 1977.
Taylor, Gordon Rattray. *The Science of Life*. New York: McGraw-Hill, 1963.

HOW DID THE MCDONALD'S DRIVE-IN RESTAURANT IN SAN BERNARDINO BECOME AN INTERNATIONAL CHAIN OF 7,000 OUTLETS?

Business Week. May 4, 1981.
Forbes. Jan. 15, 1973.
Furlong, William Barry. "Ray Kroc: Burger Master." *The Saturday Evening Post*. Mar. 1981.
McDonald's Corporation Annual Report 1981 and educational materials.

HOW DID WILLIE SUTTON DECIDE TO ROB BANKS? AND HOW DID HE DO IT?

New York Times, The. Feb. 19, 1952; Mar. 9 and 27, 1952; May 3, 1952.
New York World-Telegram. Feb. 25, 1952; Mar. 20, 1952.
Sutton, Willie, with Edward Linn. *Where the Money Was*. New York: Viking Press, 1976.

HOW DID SHIRLEY TEMPLE GET A CONTRACT FOR LIFE INSURANCE AT THE AGE OF SEVEN?

Robertson, Patrick. *Movie Facts and Feats: A Guinness Record Book*. New York: Sterling Publishing Co., 1980.

HOW DID W. T. GRANT, INC., GO BROKE?

Loving, Rush, Jr. "W. T. Grant's Last Days—As Seen from Store 1192." *Fortune*. Apr. 1976.
New York Times, The. Jan. 27, 1976; Mar. 13 and 14, 1976; Feb. 4, 1977.
Time. Mar. 22, 1976.
Women's Wear Daily. Feb. 3 and 4, 1977.

HOW DID THEY DISCOVER NEPTUNE?

Hawkins, Gerald. *Splendor in the Sky*. New York: Harper & Row, 1969.
Pickering, James Sayre. *1001 Questions Answered About Astronomy*. New York: Dodd, Mead, 1966.
Ronan, Colin. *The Astronomers*. New York: Hill & Wang, 1964.
Ronan, Colin. *The Astronomers Royal*. New York: Doubleday & Co., 1969.
Ronan, Colin. *Discovering the Universe: A History of Astronomy*. New York: Basic Books, 1971.

HOW DID THEY DRAW THE MASON-DIXON LINE A CENTURY BEFORE THE CIVIL WAR?

Encyclopedia Americana.
Tuleja, Tad. *Fabulous Fallacies*. New York: Harmony Books, 1982.

HOW DID THE WHOOPING CRANE LEARN TO REPRODUCE IN CAPTIVITY?

Discover. Aug. 1982.
Life. Aug. 1980.
McNulty, Faith. *The Whooping Crane: The Bird That Defies Extinction*. New York: E. P. Dutton, 1966.
New York Times, The. Mar. 23, 1980; Aug. 30, 1981.

HOW DID LOUIS XIV DINE?

Bradby, G.F. *The Great Days of Versailes*. New York: Charles Scribner's Sons, 1906.

Braudel, Fernand. *The Structures of Everyday Life*. New York: Harper & Row, 1981.

de Gramont, Sanche, ed. *The Age of Magnificence: The Memoirs of the Duc de Saint-Simon*. New York: G. P. Putnam's Sons, 1963.

Trager, James. *The Enriched, Fortified, Concentrated, Country-Fresh, Lip-Smacking, Finger-Licking, International, Unexpurgated Foodbook*. New York: Grossman Publishers, 1970.

HOW DID THEY DECIDE THE LENGTH OF A MILE?

Encyclopaedia Britannica.

Johnstone, William D. *For Good Measure: A Complete Compendium of International Weights and Measures*. New York: Holt, Rinehart & Winston, 1975.

Zupko, Ronald Edward. *A Dictionary of English Weights and Measures from Anglo Saxon Times to the Nineteenth Century*. Madison, Wis.: University of Wisconsin Press, 1968.

HOW DID THEY BUILD THE GREAT PYRAMID AT GIZA?

Breasted, James Henry. *A History of Egypt*. New York: Charles Scribner's Sons, 1916.

Dunham, Dows. "Building a Pyramid." Boston: Museum of Fine Arts, 1972.

Edwards, I.E.S. *The Pyramids of Egypt*. New York: Viking Press, 1947.

Fakhry, Ahmed. *The Pyramids*. Chicago: University of Chicago Press, 1961.

Lucas, A. *Ancient Egyptian Materials and Industries*. London: Edward Arnold & Co., 1926.

Tompkins, Peter. *Secrets of the Great Pyramid*. New York: Harper & Row, 1971.

HOW DID BAUDELAIRE'S *LES FLEURS DU MAL* GET REVIEWED WHEN IT WAS PUBLISHED IN 1857?

de Jong, Alex. *Baudelaire: Prince of Clouds*. New York: Paddington Press, 1976.

HOW DID THEY CONTROL THE POPULATION BEFORE THE PILL?

Tannahill, Reay. *Sex in History*. New York: Stein & Day, 1982.

HOW DID THEY SET THE PRICE OF THE LOUISIANA PURCHASE?

Blum, John M., Bruce Catton, Edmund S. Morgan, Arthur M. Schlesinger, Jr., Kenneth M. Stampp, and C. Vann Woodward. *The National Experience*. New York: Harcourt, Brace & World, 1968.

HOW DID THEY LEARN THE *TITANIC* WAS SINKING?

Lord, Walter. *A Night to Remember*. New York: Henry Holt & Co., 1955.

HOW DID BACH GET THE JOB OF CANTOR AT ST. THOMAS'S CHURCH IN LEIPZIG?

Geiringer, Karl. *J. S. Bach: The Culmination of an Era*. New York: Oxford University Press, 1966.

HOW DID THEY EXCAVATE DELPHI?

Michaelis, A. *A Century of Archaeological Discoveries*. London: John Murray, 1908.
Poulsen, Frederik. *Delphi*. London: Gyldendal, 1920.
Walker, Alan. *Delphi*. Athens: Lycabettus Press, 1977.

HOW DID CHARLIE CHAPLIN MAKE UP HIS TRAMP COSTUME?

Robertson, Patrick. *Movie Facts and Feats: A Guinness Record Book*. New York: Sterling Publishing Co., 1980.

HOW DID THE UNITED STATES BUY ALASKA?

Clark, Henry W. *The History of Alaska*. New York: Macmillan Co., 1930.

Gruening, Ernest Henry. *The State of Alaska*. New York: Random House, 1954.

Sherwood, Morgan B., ed. *Alaska and Its History*. Seattle: University of Washington Press, 1967.

HOW DID JOHN QUINCY ADAMS MEET THE PRESS?

Boller, Paul F., Jr. *Presidential Anecdotes*. New York: Oxford University Press, 1981.

Colman, Edna M. *Seventy-five Years of White House Gossip*. New York: Doubleday, Page & Co., 1926.

HOW DID HOWARD HUGHES FIND HIS WOMEN?

Wallace, Irving, Amy Wallace, David Wallechinsky, and Sylvia Wallace. *The Intimate Sex Lives of Famous People*. New York: Dell Publishing Co., 1981.

HOW DID THEY KNOW THERE WAS AN EL DORADO?

Connell, Evan S. *A Long Desire*. New York: Holt, Rinehart & Winston, 1979.

HOW DID THEY KILL TROTSKY?

Payne, Robert. *The Life and Death of Trotsky*. New York: McGraw-Hill, 1977.

HOW DID THEY DECIDE HOW TALL TO MAKE THE EMPIRE STATE BUILDING?

"Bicentennial and Civil Engineers" (special issue). *Civil Engineering*. July 1976.

Goldman, Jonathan. *The Empire State Building Book*. New York: St. Martin's Press, 1980.

Helmsley-Spear. Empire State Building. *The Empire State Building, 1931–1981*.

James, Theodore, Jr. *The Empire State Building*. New York: Harper & Row, 1975.
Life. Dec. 10, 1971.
New York Times, The. May 9, 1971.
Real Estate Forum. July 1981.

HOW DID THEY INVENT THE POTATO CHIP?

Trager, John. *The Enriched, Fortified, Concentrated, Country-Fresh, Lip-Smacking, Finger-Licking, International, Unexpurgated Foodbook*. New York: Grossman Publishers, 1970.

HOW DID DYLAN THOMAS DIE?

Ferris, Paul. *Dylan Thomas*. New York: Dial Press, 1977.
Thomas, Dylan. *Selected Letters*. Ed. Constantine FitzGibbon. New York: New Directions, 1966.

HOW DID THE PILGRIMS LEARN TO GROW CORN?

Trager, John. *The Enriched, Fortified, Concentrated, Country-Fresh, Lip-Smacking, Finger-Licking, International, Unexpurgated Foodbook*. New York: Grossman Publishers, 1970.

HOW DID HIROHITO COME TO RENOUNCE HIS DIVINITY?

MacArthur, Douglas. *Reminiscences*. New York: McGraw-Hill, 1964.
Mosley, Leonard. *Hirohito: Emperor of Japan*. New Jersey: Prentice-Hall, 1966.

HOW DID THEY FIND A RIVER IN THE SAHARA?

Time. Dec. 6, 1982.

HOW DID THEY MAKE WARREN HARDING THE REPUBLICAN PRESIDENTIAL NOMINEE IN 1920?

Baltimore Evening Sun. Oct. 18, 1920.
Downes, Randolph C. *The Rise of Warren Gamaliel Harding,*

1865–1920. Columbus, Ohio: Ohio State University Press, 1970.
Murray, Robert K. *The Harding Era: Warren G. Harding and His Administration*. Minneapolis: University of Minneapolis Press, 1969.

HOW DID THEY DESIGN THE MODEL T?

Ford Motor Co. Educational Affairs Department. *The Evolution of Mass Production* and *The Model T*.
Sorensen, Charles, with Samuel T. Williamson. *My Forty Years with Ford*. New York: W. W. Norton, 1956.

HOW DID THEY DISCOVER TROY?

Blegen, Carl W. *Troy and the Trojans*. London: Thames & Hudson, 1963.
Encyclopedia Americana.
Eydoux, Henri-Paul. *In Search of Lost Worlds*. London: Hamlyn, 1972.
Fagan, Brian M. *Quest for the Past*. Reading, Mass.: Addison-Wesley, 1978.

HOW DID SUPERMAN FLY?

Brosnan, John. *Movie Magic: The Story of Special Effects in the Cinema*. New York: St. Martin's Press, 1974.
Culhane, John. *Special Effects in the Movies*. New York: Ballantine Books, 1981.

HOW DID THEY FIRST MEASURE THE WEIGHT OF THE EARTH'S ATMOSPHERE?

Whipple, A.B.C. "Storms the Angry Gods Sent Are Now Science's Quarry." *Smithsonian*. Sept. 1982.

HOW DID NEANDERTHAL MAN FIND FOOD AND SHELTER?

Leakey, Richard E. *The Making of Mankind*. New York: E. P. Dutton, 1981.
Solecki, Ralph S. *Shanidar*. New York: Alfred A. Knopf, 1971.

HOW DID 3M TAPE BECOME SCOTCH?

Wallechinsky, David, and Irving Wallace. *The People's Almanac #1*. New York: Doubleday & Co., 1975.

HOW DID THEY DISCOVER THE HOPE DIAMOND?

Desautels, Paul E. *The Gem Kingdom*. New York: Random House, 1971.

Patch, Susanne Steinem. *Blue Mystery: The Story of the Hope Diamond*. Washington, D.C.: Smithsonian Press, 1975.

Whitlock, Herbert P. *The Story of Gems*. New York: Lee Furman, 1936.

HOW DID THEY DECIDE TO PUBLISH *TIME* MAGAZINE?

Elson, Robert T. *Time Inc.: The Intimate History of a Publishing Enterprise, 1923–1941*. Ed. Duncan Norton-Taylor. New York: Atheneum, 1968.

HOW DID THEY FIND THE POEMS OF EMILY DICKINSON?

Dickinson, Emily. *Selected Poems of Emily Dickinson*. New York: Random House, Modern Library, 1948.

Taggard, Genevieve. *The Life and Mind of Emily Dickinson*. New York: Alfred A. Knopf, 1930.

HOW DID LAUREN BACALL MEET HUMPHREY BOGART?

Bacall, Lauren. *Lauren Bacall by Myself*. New York: Alfred A. Knopf, 1978.

Greenberger, Howard. *Bogey's Baby*. New York: Cornerstone Library, 1976.

HOW DID THEY PUT AL CAPONE IN JAIL?

Allsop, Kenneth. *The Bootleggers and Their Era*. New York: Doubleday & Co., 1961.

Pasley, Fred D. *Al Capone: The Biography of a Self-made Man*. New York: Garden City Publishing Co., 1930.

HOW DID THEY MAKE THE FIRST ACCURATE MAP OF THE WORLD?

Birch, T.W. *Maps*. 2nd ed. London: Oxford University Press, 1964.

Brown, Lloyd A. *Map Making*. Boston: Little, Brown & Co., 1960.

Brown, Lloyd A. *The Story of Maps*. Boston: Little, Brown & Co., 1950.

HOW DID THEY NAME THE HOT DOG?

Oscar Mayer Foods Corporation. Consumer affairs information.

Trager, John. *The Enriched, Fortified, Concentrated, Country-Fresh, Lip-Smacking, Finger-Licking, International, Unexpurgated Foodbook*. New York: Grossman Publishers, 1970.

HOW DID THEY UNIONIZE THE FORD MOTOR COMPANY?

Nevins, Allan, and Frank Ernest Hill. *Ford: Decline and Rebirth, 1933–1962*. New York: Charles Scribner's Sons, 1963.

HOW DID THE INDIANS START SCALPING?

American Heritage Book of the Indian. New York: American Heritage Publishing, 1961.

Tuleja, Tad. *Fabulous Fallacies*. New York: Harmony Books, 1982.

HOW DID STRADIVARI MAKE HIS VIOLINS?

Fetis, F.J. *Anthony Stradivari: Celebrated Violin Maker*. London: William Reeves, 1864.

Goodkind, Herbert K. *Violin Iconography of Antonio Stradivari, 1644–1737*. Larchmont, N.Y.: Published by the author, 1972.

Henley, W. *Antonio Stradivari: His Life and Instruments*. Sussex, England: Amati Publishers, 1961.

Hill, W. Henry, Arthur F. Hill, and Alfred E. Hill. *Antonio Stradivari: His Life and Work (1644–1737)*. New York: Dover Publications, 1963 (reprint of work published in 1902).

Nelson, Sheila M. *The Violin and Viola*. New York: W. W. Norton, 1972.
Sadie, Stanley, ed. *The New Grove Dictionary of Music and Musicians*. Vol. 18. New York: Macmillan Co., 1980.
Wechsberg, Joseph. *The Glory of the Violin*. New York: Viking Press, 1972.

HOW DID THEY DETERMINE THE AGE OF THE EARTH?

Jastrow, Robert. *Red Giants and White Dwarfs: The Evolution of Stars, Planets, and Life*. New York: Harper & Row, 1967.
Ronan, Colin. *Discovering the Universe: A History of Astronomy*. New York: Basic Books, 1971.

HOW DID MAE WEST'S SECRET HUSBAND LIKE MARRIED LIFE?

Eells, George, and Stanley Musgrove. *Mae West: A Biography*. New York: William Morrow & Co., 1982.

HOW DID THEY START THE CHICAGO FIRE OF 1871?

Encyclopedia Americana.
Tuleja, Tad. *Fabulous Fallacies*. New York: Harmony Books, 1982.

HOW DID THEY FIRST LEARN TO MAKE COFFEE?

Encyclopedia Britanica.
International Coffee Organization. *United States of America, Coffee Drinking Study, Winter 1982*. London 1982.
Uribe Compuzano, Adrés. *Brown Gold: The Amazing Story of Coffee*. New York: Random House, 1954.

HOW DID NAPOLEON ESCAPE FROM ELBA?

Castelot, André. *Napoleon*. New York: Harper & Row, 1967.
Kircheisen, F.M. *Napoleon*. New York: Harcourt, Brace & Co., 1932.
Thompson, J.M. *Napoleon Bonaparte*. New York: Oxford University Press, 1952.

HOW DID LOUISA MAY ALCOTT WRITE *LITTLE WOMEN*?

Wallechinsky, David, and Irving Wallace. *The People's Almanac #3*. New York: William Morrow & Co., 1981.

HOW DID MEN DECIDE TO WEAR NECKTIES?

Kybalova, Ludmila, Olga Herbenova, and Milena Lamarova. *The Pictorial Encyclopedia of Fashion*. New York: Crown Publishers, 1968.

HOW DID BEETHOVEN COMPOSE WHEN HE WAS DEAF?

Anderson, Emily. ed. *Letters of Beethoven*. New York: St. Martin's Press, 1961.

Marek, George Richard. *Beethoven: Biography of a Genius*. New York: Funk and Wagnalls, 1969.

Schindler, Anton Felix. *Beethoven as I Knew Him*. Ed. Donald W. MacArdle. Chapel Hill, N.C.: University of North Carolina Press, 1966.

HOW DID THEY BUILD THE GREAT WALL OF CHINA?

Encyclopaedia Britannica.

Hucker, Charles O. *China's Imperial Past: An Introduction to Chinese History and Culture*. Stanford, Calif.: Stanford University Press, 1975.

New York Times, The. Mar. 20, 1983.

Zewen, Luo, Dai Wenbao, Dick Wilson, Jean-Pierre Drege, and Hubert Delahaye. *The Great Wall*. New York: McGraw-Hill, 1981.

HOW DID GUTENBERG PRINT THE BIBLE?

Blumenthal, Joseph. *Art of the Printed Book, 1455–1955: Masterpieces of Typography Through Five Centuries from the Collections of the Pierpont Morgan Library, New York*. Boston: David R. Godine. New York: Pierpont Morgan Library, 1973.

Chappell, Warren. *A Short History of the Printed Word*. New York: Alfred A. Knopf, 1970.
Geck, Elisabeth. *Johann Gutenberg: From Lead Letter to the Computer, 1468–1968*. Bad Godesberg, West Germany: Inter Nationes, 1968.

HOW DID THEY DEVELOP THE PARACHUTE?

Encyclopaedia Britannica.

HOW DID HARPO LEARN TO PLAY THE HARP?

Arce, Hector. *Groucho*. New York: G. P. Putnam's Sons, 1979.
Marx, Harpo, with Rowland Barber. *Harpo Speaks!* London: Victor Gollancz, 1961.
Zimmerman, Paul D., and Burt Goldblatt. *The Marx Brothers at the Movies*. New York: G. P. Putnam's Sons, 1968.

HOW DID FDR COME TO APPOINT THE FIRST WOMAN TO A PRESIDENT'S CABINET?

Boller, Paul F., Jr. *Presidential Anecdotes*. New York: Oxford University Press, 1981.
Lash, Joseph P. *Eleanor and Franklin: The Story of Their Relationship Based on Eleanor Roosevelt's Private Papers*. New York: W. W. Norton, 1971.

HOW DID THEY SPLIT THE ATOM?

Campbell, John W. *The Atomic Story*. New York: Henry Holt & Co., 1947.
Compton, Arthur Holly. *Atomic Quest*. New York: Oxford University Press, 1956.
Laurence, William. *Men and Atoms*. New York: Simon & Schuster, 1959.
Snow, C.P. *The Physicists*. Boston: Little, Brown & Co., 1981.

HOW DID THEY NAME AMERICA?

Wallechinsky, David, and Irving Wallace. *The People's Almanac*. New York: Doubleday & Co., 1975.

HOW DID THEY INVENT CIGARETTES?

Encyclopaedia Britannica.

HOW DID THEY KILL MARIE ANTOINETTE?

Anthony, Katharine. *Marie Antoinette*. New York: Alfred A. Knopf, 1933.

Hearsey, John. *Marie Antoinette*. New York: E. P. Dutton, 1973.

Palache, John Garber. *Marie Antoinette: The Player Queen*. London: Longmans, Green & Co., 1929.

HOW DID THEY KNOW THE SIZE OF THE EARTH OVER 1,700 YEARS BEFORE ANYONE SAILED AROUND IT?

Mitton, Simon, ed. *The Cambridge Encyclopedia of Astronomy*. New York: Crown Publishers, 1977.

HOW DID THEY DETERMINE THE MEMBERSHIP OF THE U.S. SENATE?

Blum, John M., Bruce Catton, Edmund S. Morgan, Arthur M. Schlesinger, Jr., Kenneth M. Stampp, and C. Vann Woodward. *The National Experience*. New York: Harcourt, Brace & World, 1968.

HOW DID HANNIBAL CROSS THE ALPS?

de Beer, Sir Gavin. *Hannibal: The Struggle for Power in the Mediterranean*. London: Thames & Hudson, 1969.

HOW DID THEY DISCOVER RADIUM?

Curie, Eve. *Madame Curie*. New York: Doubleday, Doran & Co., 1937.

Curie, Marie. *Pierre Curie*. New York: Macmillan Co., 1923.

Reid, Robert. *Marie Curie*. New York: E. P. Dutton and Saturday Review Press, 1974.

INDEX

A

B

C

D

E

F

G

H

I

J

L

M

N

O

P

R

S

T

Also available from Quill

How Do They Do That?
Wonders of the Modern World Explained
Caroline Sutton

How do bees build honeycombs? How are tunnels dug underwater? This book entertainingly unravels some of the modern world's greatest puzzles—a browser's delight. 0-688-01111-X

Big Secrets
The Uncensored Truth About All Sorts of Stuff You Are Never Supposed to Know
William Poundstone

What's really in Coca-Cola? Is there sand in Crest toothpaste? How does Doug Henning do his illusions? How can you beat a lie detector? American institutions love to keep secrets—secret formulas, secret sauces, secret handshakes—and *Big Secrets* spills the beans with wit and humor. 0-688-04830-7

One Night Stands with American History
Odd, Amusing, and Little-Known Incidents
Richard Shenkman and Kurt Reiger

A treasure trove of obscure stories about famous Americans, peculiar incidents in our history, and exposés of our cultural idiosyncrasies. 0-688-01399-6

AVAILABLE AT YOUR LOCAL BOOKSTORE